Studies in Advanced Mathematics

Titles Included in the Series

Steven R. Bell, The Cauchy Transform, Potential Theory, and Conformal Mapping
John J. Benedetto, Harmonic Analysis and Applications
John J. Benedetto and Michael W. Frazier, Wavelets: Mathematics and Applications
Albert Boggess, CR Manifolds and the Tangential Cauchy–Riemann Complex
Goong Chen and Jianxin Zhou, Vibration and Damping in Distributed Systems, Vol. 1: Analysis, Estimation, Attenuation, and Design. Vol. 2: WKB and Wave Methods, Visualization, and Experimentation
Carl C. Cowen and Barbara D. MacCluer, Composition Operators on Spaces of Analytic Functions
John P. D'Angelo, Several Complex Variables and the Geometry of Real Hypersurfaces
Lawrence C. Evans and Ronald F. Gariepy, Measure Theory and Fine Properties of Functions
Gerald B. Folland, A Course in Abstract Harmonic Analysis
José García-Cuerva, Eugenio Hernández, Fernando Soria, and José-Luis Torrea, Fourier Analysis and Partial Differential Equations
Peter B. Gilkey, Invariance Theory, the Heat Equation, and the Atiyah-Singer Index Theorem, 2nd Edition
Alfred Gray, Modern Differential Geometry of Curves and Surfaces with Mathematica, 2nd Edition
Eugenio Hernández and Guido Weiss, A First Course on Wavelets
Steven G. Krantz, Partial Differential Equations and Complex Analysis
Steven G. Krantz, Real Analysis and Foundations
Kenneth L. Kuttler, Modern Analysis
Clark Robinson, Dynamical Systems: Stability, Symbolic Dynamics, and Chaos, 2nd Edition
John Ryan, Clifford Algebras in Analysis and Related Topics
Xavier Saint Raymond, Elementary Introduction to the Theory of Pseudodifferential Operators
Robert Strichartz, A Guide to Distribution Theory and Fourier Transforms
André Unterberger and Harald Upmeier, Pseudodifferential Analysis on Symmetric Cones
James S. Walker, Fast Fourier Transforms, 2nd Edition
Gilbert G. Walter, Wavelets and Other Orthogonal Systems with Applications
Kehe Zhu, An Introduction to Operator Algebras

A Primer on WAVELETS and their Scientific Applications

James S. Walker
University of Wisconsin-Eau Claire

CHAPMAN & HALL/CRC

Boca Raton London New York Washington, D.C.

Library of Congress Cataloging-in-Publication Data

Walker, James S.
A primer on wavelets and their scientific applications / James S. Walker.
p. cm.—(Studies in advanced mathematics)
Includes bibliographical references (p. -) and index.
ISBN 0-8493-8276-9 (alk. paper)
1. Wavelets (Mathematics) I. Title. II. Series.
QA403.3.W33 1999
515′.2433—dc21 98-55724
CIP

Direct all inquiries to CRC Press LLC, 2000 N.W. Corporate Blvd., Boca Raton, Florida 33431, or visit our Web site at www.crcpress.com

International Standard Book Number 0-8493-8276-9
Library of Congress Card Number 98-55724
Printed in the United States of America 5 6 7 8 9 0
Printed on acid-free paper

Preface

"Wavelet theory" is the result of a multidisciplinary effort that brought together mathematicians, physicists and engineers...this connection has created a flow of ideas that goes well beyond the construction of new bases or transforms.

Stéphane Mallat[1]

The past decade has witnessed an explosion of activity in wavelet analysis. Thousands of research papers have been published on the theory and applications of wavelets. Wavelets provide a powerful and remarkably flexible set of tools for handling fundamental problems in science and engineering.

For an idea of the wide range of problems that are being solved using wavelets, here is a list of some of the problems discussed in this book:

- *Audio denoising:* Long distance telephone messages often contain significant amounts of noise. How do we remove this noise in order to clarify the messages?
- *Signal compression:* The efficient transmission of large amounts of data, over the Internet for example, requires some kind of compression. Are there ways we can compress this data as much as possible without losing significant information?
- *Object detection:* What methods can we use to pick out a small image, say of an aircraft, from the midst of a larger more complicated image?
- *Fingerprint compression:* The FBI has 25 million fingerprint records. If these fingerprint records were digitized without any compression,

[1]Mallat's quote is from [MAL].

they would gobble up 250 *trillion* bytes of storage capacity. Is there a way to compress these records to a manageable size, without losing any significant details in the fingerprints?

- *Image denoising:* Images formed by electron microscopes and by optical lasers are often contaminated by large amounts of unwanted clutter (referred to as *noise*). Can this noise be removed in order to clarify the image?

- *Image enhancement:* When an optical microscope image is recorded, it often suffers from blurring. How can the appearance of the objects in these images be sharpened?

- *Image recognition:* How do humans recognize faces? Can we teach machines to do it?

- *Diagnosing heart trouble:* Is there a way to detect abnormal heartbeats, hidden within a complicated electrocardiogram?

- *Speech recognition:* What factors distinguish consonants from vowels? How do humans recognize different voices?

All of these problems can be tackled using wavelets. We will show how during the course of this book.

In Chapter 1 we introduce the simplest wavelets, the Haar wavelets. We also introduce many of the basic concepts—wavelet transforms, energy conservation and compaction, multiresolution analysis, compression and denoising—that will be used in the remainder of the book. For this reason, we devote more pages to the theory of Haar wavelets than perhaps they deserve alone; keep in mind that this material will be amplified and generalized throughout the remainder of the book.

Chapter 2 is the heart of the book. In this chapter we describe the Daubechies wavelets, which have played a key role in the explosion of activity in wavelet analysis. After a simple introduction to their mathematical properties we then describe several applications of these wavelets. First, we explain in detail how they can be used to compress audio signals—this application is vital to the fields of telephony and telecommunications. Second, we describe how a method known as *thresholding* provides a powerful technique for removing random noise (static) from audio signals. Removing random noise is a fundamental necessity when dealing with all kinds of data in science and engineering. The threshold method, which is analogous to how our nervous system responds only to inputs above certain thresholds, provides a nearly optimal method for removing random noise. Besides random noise, Daubechies wavelets can also be used to remove isolated "pop-noise" from audio.

Wavelet analysis can also be applied to images. We shall examine compression of images, including fingerprint compression, and denoising of images. The image denoising examples that we examine include some examples motivated by *magnetic resonance imaging* (MRI) and laser imaging.

Chapter 2 concludes with some examples from image processing. We discuss edge detection, and the sharpening of blurred images, and an example from computer vision where wavelet methods can be used to enormously increase the speed of identification of an image.

Chapter 3 relates wavelet analysis to frequency analysis. Frequency analysis, also known as *Fourier analysis,* has long been one of the cornerstones of the mathematics of science and engineering. We shall briefly describe how wavelets are characterized in terms of their effects on the frequency content of signals. One application that we discuss is object identification—locating a small object within a complicated scene—where wavelet analysis in concert with Fourier analysis provides a powerful approach.

In the final chapter we deal with some extensions which reach beyond the fundamentals of wavelets. We describe a generalization of wavelet transforms known as *wavelet packet transforms.* We apply these wavelet packet transforms to compression of audio signals, images, and fingerprints. Then we turn to the subject of *continuous wavelet transforms*, as they are implemented in a discrete form on a computer. Continuous wavelet transforms are widely used in seismology and have also been used very effectively for analyzing speech and electrocardiograms.

The goal of this primer is to guide the reader through the main ideas of wavelet analysis to facilitate a knowledgeable reading of the present research literature, especially in the applied fields of audio and image processing and biomedicine. Although there are several excellent books on the theory of wavelets, these books are focused on the construction of wavelets and their mathematical properties. Furthermore, they are all written at a graduate school level of mathematical and/or engineering expertise. There is a real need for a simple introduction, a *primer,* which uses only elementary algebra and a smidgen of calculus to explain the underlying ideas behind wavelet analysis, and devotes the majority of its pages to explaining how these underlying ideas can be applied to solve significant problems in audio and image processing and in biology and medicine.

To keep the mathematics simple, we focus on the discrete theory—technically known as *subband coding.* It is in the continuous theory of wavelet analysis where the most difficult mathematics lies; yet when this continuous theory is applied it is almost always converted into the discrete approach that we describe in this primer. Focusing on the discrete case will allow us to concentrate on the applications of wavelet analysis while at the same time keeping the mathematics under control. On the rare occasions when we need to use more advanced mathematics, we shall mark these discussions off from the main text by putting them into subsections that are marked by asterisks in their titles. An effort has been made to ensure that subsequent discussions do not rely on this more advanced material.

Without question the best way, perhaps the only way, to learn about applications of wavelets is to experiment with making such applications. This experimentation is typically done on a computer. In order to simplify this computer experimentation, I have created software, called FAWAV, which can be downloaded over the Internet—see Appendix A. FAWAV runs under WINDOWS™ 95, 98, and NT 4.0, and *requires no programming to use.* Further details about FAWAV can be found in Appendix A; suffice it for now to say that it is designed to allow the reader to duplicate all of the applications described in this primer and to experiment with other ideas.

This primer is only a first introduction to wavelets and their scientific applications. For that reason we limit ourselves to describing what are known technically as periodic orthogonal wavelets. Other types of wavelets—biorthogonal wavelets, spline wavelets, multiwavelets, etc.—are also used in applications, but we feel that these other wavelets can be understood much more easily if the periodic orthogonal ones are studied first. In the Notes and references sections that conclude each chapter, we provide the reader with ample references where further information on these other wavelets and many other topics can be found.

Acknowledgments

It is a pleasure to thank everyone who has assisted me during the writing of this book. First, let me mention my Executive Editor at CRC Press, Bob Stern, who first suggested the idea of this book to me and helped me to pursue it. My Series Editor, Steve Krantz, supplied several helpful criticisms and some much appreciated encouragement when I really needed it. Greg Smethells, a former student of mine, supplied many helpful comments on both the book manuscript and FAWAV. The students of my fall 1997 Fourier Optics course deserve recognition for putting up with some of the early versions of the image processing applications described in this book. I think their many questions and suggestions have improved my presentation of these ideas. C.L. Tondo, TeX wizard, helped me overcome several difficulties with the LaTeX system used to typeset the book. Mimi Williams saved me from committing several grammatical errors. Beeneet Kothari, Walter Reid, and Robb Sloan provided some useful opinions on writing style. At last, a big "shei shei" to my dear wife Angela (Ching-Shiow) for her meticulous proofreading, and for her understanding and support during the many long days and nights spent writing this book.

James S. Walker
Eau Claire, Wisconsin
October 28, 1998

To my wife, and the memory of my mother

Contents

Chapter 1

Haar Wavelets

The purpose of computing is insight, not numbers.

Richard W. Hamming

The purpose of computing is insight, not pictures.

Lloyd N. Trefethen[1]

A Haar wavelet is the simplest type of wavelet. In discrete form, Haar wavelets are related to a mathematical operation called the *Haar transform*. The Haar transform serves as a prototype for all other wavelet transforms. Studying the Haar transform in detail will provide a good foundation for understanding the more sophisticated wavelet transforms which we shall describe in the next chapter. In this chapter we shall describe how the Haar transform can be used for compressing audio signals and for removing noise. Our discussion of these applications will set the stage for the more powerful wavelet transforms to come and their applications to these same problems. One distinctive feature that the Haar transform enjoys is that it lends itself easily to simple hand calculations. We shall illustrate many concepts by both simple hand calculations and more involved computer computations.

1.1 The Haar transform

In this section we shall introduce the basic notions connected with the Haar transform, which we shall examine in more detail in later sections.

[1]Hamming's quote is from [HAM]. Trefethen's quote is from [TRE].

First, we need to define the type of signals that we shall be analyzing with the Haar transform.

Throughout this book we shall be working extensively with *discrete signals.* A discrete signal is a function of time with values occurring at discrete instants. Generally we shall express a discrete signal in the form $\mathbf{f} = (f_1, f_2, \ldots, f_N)$, where N is a positive even integer which we shall refer to as the *length* of $\mathbf{f}$. The *values* of $\mathbf{f}$ are the N real numbers $f_1, f_2, \ldots, f_N$. These values are typically measured values of an analog signal g, measured at the time values $t = t_1, t_2, \ldots, t_N$. That is, the values of $\mathbf{f}$ are

$$f_1 = g(t_1),\ f_2 = g(t_2),\ \ldots,\ f_N = g(t_N). \tag{1.1}$$

For simplicity, we shall assume that the increment of time that separates each pair of successive time values is always the same. We shall use the phrase *equally spaced sample values*, or just *sample values,* when the discrete signal has its values defined in this way. An important example of sample values is the set of data values stored in a computer audio file, such as a `.wav` file. Another example is the sound intensity values recorded on a compact disc. A non-audio example, where the analog signal g is not a sound signal, is a digitized electrocardiogram.

Like all wavelet transforms, the Haar transform decomposes a discrete signal into two subsignals of half its length. One subsignal is a running average or *trend;* the other subsignal is a running difference or *fluctuation.*

Let's begin by examining the trend subsignal. The *first trend* subsignal, $\mathbf{a}^1 = (a_1, a_2, \ldots, a_{N/2})$, for the signal $\mathbf{f}$ is computed by taking a running average in the following way. Its first value, a_1, is computed by taking the average of the first pair of values of $\mathbf{f}$: $(f_1 + f_2)/2$, and then multiplying it by $\sqrt{2}$. That is, $a_1 = (f_1 + f_2)/\sqrt{2}$. Similarly, its next value a_2 is computed by taking the average of the next pair of values of $\mathbf{f}$: $(f_3 + f_4)/2$, and then multiplying it by $\sqrt{2}$. That is, $a_2 = (f_3 + f_4)/\sqrt{2}$. Continuing in this way, all of the values of $\mathbf{a}^1$ are produced by taking averages of successive pairs of values of $\mathbf{f}$, and then multiplying these averages by $\sqrt{2}$. A precise formula for the values of $\mathbf{a}^1$ is

$$a_m = \frac{f_{2m-1} + f_{2m}}{\sqrt{2}}, \tag{1.2}$$

for $m = 1, 2, 3, \ldots, N/2$.

For example, suppose $\mathbf{f}$ is defined by eight values, say

$$\mathbf{f} = (4, 6, 10, 12, 8, 6, 5, 5);$$

then its first trend subsignal is $\mathbf{a}^1 = (5\sqrt{2}, 11\sqrt{2}, 7\sqrt{2}, 5\sqrt{2})$. This result can be obtained using Formula (1.2). Or it can be calculated as indicated in the following diagram:

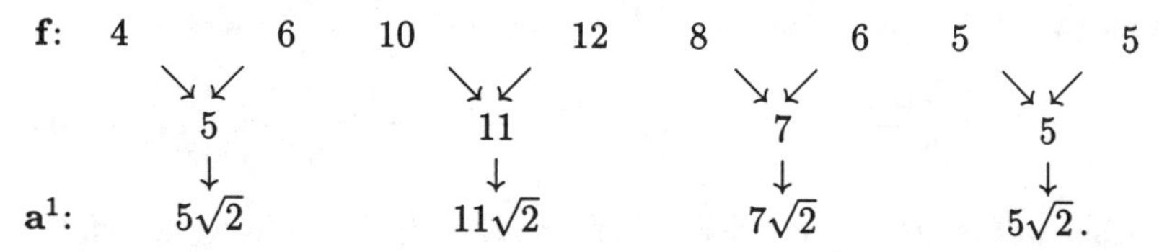

You might ask: Why perform the extra step of multiplying by $\sqrt{2}$? Why not just take averages? These questions will be answered in the next section, when we show that multiplication by $\sqrt{2}$ is needed in order to ensure that the Haar transform preserves the energy of a signal.

The other subsignal is called the *first fluctuation*. The first fluctuation of the signal $\mathbf{f}$, which is denoted by $\mathbf{d}^1 = (d_1, d_2, \ldots, d_{N/2})$, is computed by taking a running difference in the following way. Its first value, d_1, is calculated by taking half the difference of the first pair of values of $\mathbf{f}$: $(f_1 - f_2)/2$, and multiplying it by $\sqrt{2}$. That is, $d_1 = (f_1 - f_2)/\sqrt{2}$. Likewise, its next value d_2 is calculated by taking half the difference of the next pair of values of $\mathbf{f}$: $(f_3 - f_4)/2$, and multiplying it by $\sqrt{2}$. In other words, $d_2 = (f_3 - f_4)/\sqrt{2}$. Continuing in this way, all of the values of $\mathbf{d}^1$ are produced according to the following formula:

$$d_m = \frac{f_{2m-1} - f_{2m}}{\sqrt{2}}, \tag{1.3}$$

for $m = 1, 2, 3, \ldots, N/2$.

For example, for the signal $\mathbf{f} = (4, 6, 10, 12, 8, 6, 5, 5)$ considered above, its first fluctuation $\mathbf{d}^1$ is $(-\sqrt{2}, -\sqrt{2}, \sqrt{2}, 0)$. This result can be obtained using Formula (1.3), or it can be calculated as indicated in the following diagram:

$$\begin{array}{lcccccccc}
\mathbf{f}: & 4 & & 6 & 10 & & 12 & 8 & & 6 & 5 & & 5 \\
& & \searrow\swarrow & & & \searrow\swarrow & & & \searrow\swarrow & & & \searrow\swarrow & \\
& & -1 & & & -1 & & & 1 & & & 0 & \\
& & \downarrow & & & \downarrow & & & \downarrow & & & \downarrow & \\
\mathbf{d}^1: & & -\sqrt{2} & & & -\sqrt{2} & & & \sqrt{2} & & & 0. &
\end{array}$$

Haar transform, 1-level

The Haar transform is performed in several stages, or levels. The first level is the mapping $\mathbf{H}_1$ defined by

$$\mathbf{f} \overset{\mathbf{H}_1}{\longmapsto} (\mathbf{a}^1 \,|\, \mathbf{d}^1) \tag{1.4}$$

from a discrete signal $\mathbf{f}$ to its first trend $\mathbf{a}^1$ and first fluctuation $\mathbf{d}^1$. For example, we showed above that

$$(4, 6, 10, 12, 8, 6, 5, 5) \overset{\mathbf{H}_1}{\longmapsto} (5\sqrt{2}, 11\sqrt{2}, 7\sqrt{2}, 5\sqrt{2} \,|\, -\sqrt{2}, -\sqrt{2}, \sqrt{2}, 0). \tag{1.5}$$

The mapping $\mathbf{H}_1$ in (1.4) has an inverse. Its inverse maps the transform signal $(\mathbf{a}^1 \,|\, \mathbf{d}^1)$ back to the signal $\mathbf{f}$, via the following formula:

$$\mathbf{f} = \left(\frac{a_1 + d_1}{\sqrt{2}}, \frac{a_1 - d_1}{\sqrt{2}}, \ldots, \frac{a_{N/2} + d_{N/2}}{\sqrt{2}}, \frac{a_{N/2} - d_{N/2}}{\sqrt{2}} \right). \tag{1.6}$$

In other words, $f_1 = (a_1 + d_1)/\sqrt{2}$, $f_2 = (a_1 - d_1)/\sqrt{2}$, $f_3 = (a_2 + d_2)/\sqrt{2}$, $f_4 = (a_2 - d_2)/\sqrt{2}$, and so on. For instance, the following diagram shows how to invert the transformation in (1.5):

$$\begin{array}{lcccccccc}
\mathbf{a}^1: & \multicolumn{2}{c}{5\sqrt{2}} & \multicolumn{2}{c}{11\sqrt{2}} & \multicolumn{2}{c}{7\sqrt{2}} & \multicolumn{2}{c}{5\sqrt{2}} \\
\mathbf{d}^1: & \multicolumn{2}{c}{-\sqrt{2}} & \multicolumn{2}{c}{-\sqrt{2}} & \multicolumn{2}{c}{\sqrt{2}} & \multicolumn{2}{c}{0} \\
 & \swarrow & \searrow & \swarrow & \searrow & \swarrow & \searrow & \swarrow & \searrow \\
\mathbf{f}: & 4 & 6 & 10 & 12 & 8 & 6 & 5 & 5.
\end{array}$$

Let's now consider what advantages accrue from performing the Haar transformation. These advantages will be described in more detail later in this chapter, but some basic notions can be introduced now. All of these advantages stem from the following cardinal feature of the Haar transform (a feature that will be even more prominent for the Daubechies transforms described in the next chapter):

Small Fluctuations Feature. *The magnitudes of the values of the fluctuation subsignal are often significantly smaller than the magnitudes of the values of the original signal.*

For instance, for the signal $\mathbf{f} = (4, 6, 10, 12, 8, 6, 5, 5)$ considered above, its eight values have an average magnitude of 7. On the other hand, for its first fluctuation $\mathbf{d}^1 = (-\sqrt{2}, -\sqrt{2}, \sqrt{2}, 0)$, the average of its four magnitudes is $0.75\sqrt{2}$. In this case, the magnitudes of the fluctuation's values are an average of 6.6 times smaller than the magnitudes of the original signal's values. For a second example, consider the signal shown in Figure 1.1(a). This signal was generated from 1024 sample values of the function

$$g(x) = 20x^2(1 - x)^4 \cos 12\pi x$$

over the interval $[0, 1)$. In Figure 1.1(b) we show a graph of the 1-level Haar transform of this signal. The trend subsignal is graphed on the left half, over the interval $[0, 0.5)$, and the fluctuation subsignal is graphed on the right half, over the interval $[0.5, 1)$. It is clear that a large percentage of the fluctuation's values are close to 0 in magnitude, another instance of the Small Fluctuations Feature. Notice also that the trend subsignal looks like the original signal, although shrunk by half in length and expanded by a factor of $\sqrt{2}$ vertically.

The reason that the Small Fluctuations Feature is generally true is that typically we are dealing with signals whose values are samples of a continuous analog signal g with a very short time increment between the samples.

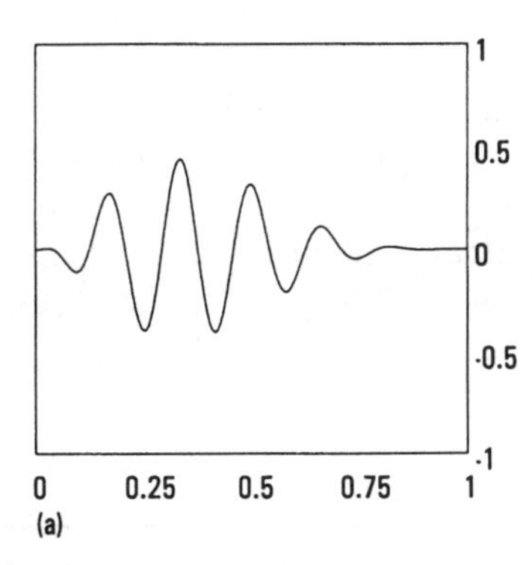

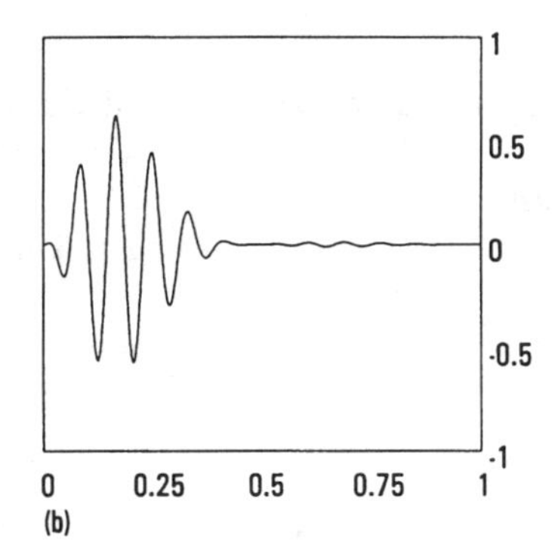

FIGURE 1.1
(a) Signal, (b) Haar transform, 1-level.

In other words, the equations in (1.1) hold with a small value of the time increment $h = t_{k+1} - t_k$ for each $k = 1, 2, \ldots, N-1$. If the time increment is small enough, then successive values $f_{2m-1} = g(t_{2m-1})$ and $f_{2m} = g(t_{2m})$ of the signal $\mathbf{f}$ will be close to each other due to the continuity of g. Consequently, the fluctuation values for the Haar transform satisfy

$$d_m = \frac{g(t_{2m-1}) - g(t_{2m})}{\sqrt{2}} \approx 0.$$

This explains why the Small Fluctuations Feature is generally true for the Haar transform. A similar analysis shows why the trend subsignal has a graph that is similar in appearance to the first trend. If g is continuous and the time increment is very small, then $g(t_{2m-1})$ and $g(t_{2m})$ will be close to each other. Expressing this fact as an approximation, $g(t_{2m-1}) \approx g(t_{2m})$, we obtain the following approximation for each value a_m of the trend subsignal

$$a_m \approx \sqrt{2}\, g(t_{2m}).$$

This equation shows that $\mathbf{a}^1$ is approximately the same as sample values of $\sqrt{2}\, g(x)$ for $x = t_2, t_4, \ldots, t_N$. In other words, it shows that the graph of the first trend subsignal is similar in appearance to the graph of g, as we pointed out above in regard to the signal in Figure 1.1(a). We shall examine these points in more detail in the next chapter when we discuss other wavelet transforms.

One of the reasons that the Small Fluctuations Feature is important is that it has applications to *signal compression*. By compressing a signal we mean transmitting its values, or approximations of its values, by using a smaller number of bits. For example, if we were only to transmit the trend subsignal for the signal shown in Figure 1.1(a) and then perform Haar transform inversion (treating the fluctuation's values as all zeros), then we would obtain an approximation of the original signal. Since the length of the trend subsignal is half the length of the original signal, this would achieve 50% compression. We shall discuss compression in more detail in Section 1.5.

Once we have performed a 1-level Haar transform, then it is easy to repeat the process and perform multiple-level Haar transforms. We shall discuss this in the next section.

1.2 Conservation and compaction of energy

In the previous section we defined the 1-level Haar transform. In this section we shall discuss its two most important properties: (1) It conserves the energies of signals; (2) It performs a compaction of the energy of signals. We shall also complete our definition of the Haar transform by showing how to extend its definition to multiple levels.

Conservation of energy

An important property of the Haar transform is that it *conserves the energies of signals.* By the *energy* of a signal $\mathbf{f}$ we mean the sum of the squares of its values. That is, the energy $\mathcal{E}_{\mathbf{f}}$ of a signal $\mathbf{f}$ is defined by

$$\mathcal{E}_{\mathbf{f}} = f_1^2 + f_2^2 + \cdots + f_N^2. \tag{1.7}$$

We shall provide some explanation for why we give the name Energy to this quantity $\mathcal{E}_{\mathbf{f}}$ in a moment. First, however, let's look at an example of calculating energy. Suppose $\mathbf{f} = (4, 6, 10, 12, 8, 6, 5, 5)$ is the signal considered in Section 1.1. Then $\mathcal{E}_{\mathbf{f}}$ is calculated as follows:

$$\mathcal{E}_{\mathbf{f}} = 4^2 + 6^2 + \cdots + 5^2 = 446.$$

So the energy of $\mathbf{f}$ is 446. Furthermore, using the values for its 1-level Haar transform $(\mathbf{a}^1 \,|\, \mathbf{d}^1) = (5\sqrt{2}, 11\sqrt{2}, 7\sqrt{2}, 5\sqrt{2} \,|\, -\sqrt{2}, -\sqrt{2}, \sqrt{2}, 0)$, we find that

$$\mathcal{E}_{(\mathbf{a}^1 \,|\, \mathbf{d}^1)} = 25 \cdot 2 + 121 \cdot 2 + \cdots + 2 + 0 = 446$$

as well. Thus the 1-level Haar transform has kept the energy constant. In fact, this is true in general:

Conservation of Energy. *The 1-level Haar transform conserves energy, i.e.,* $\mathcal{E}_{(\mathbf{a}^1 \,|\, \mathbf{d}^1)} = \mathcal{E}_{\mathbf{f}}$ *for every signal* $\mathbf{f}$.

We will explain why this Conservation of Energy property is true for all signals at the end of this section.

Before we go any further, we should say something about why we have given the name Energy to the quantity $\mathcal{E}_{\mathbf{f}}$. The reason is that sums of squares frequently appear in physics when various types of energy are calculated. For instance, if a particle of mass m has a velocity of $\mathbf{v} = (v_1, v_2, v_3)$, then its kinetic energy is $(m/2)(v_1^2 + v_2^2 + v_3^2)$. Hence its kinetic energy is proportional to $v_1^2 + v_2^2 + v_3^2 = \mathcal{E}_{\mathbf{v}}$. Ignoring the constant of proportionality, $m/2$, we obtain the quantity $\mathcal{E}_{\mathbf{v}}$ which we call the energy of $\mathbf{v}$.

While Conservation of Energy is certainly an important property, it is even more important to consider how the Haar transform redistributes the energy in a signal by compressing most of the energy into the trend subsignal. For example, for the signal $\mathbf{f} = (4, 6, 10, 12, 8, 6, 5, 5)$ we found in Section 1.1 that its trend $\mathbf{a}^1$ equals $(5\sqrt{2}, 11\sqrt{2}, 7\sqrt{2}, 5\sqrt{2})$. Therefore, the energy of $\mathbf{a}^1$ is

$$\mathcal{E}_{\mathbf{a}^1} = 25 \cdot 2 + 121 \cdot 2 + 49 \cdot 2 + 25 \cdot 2 = 440.$$

On the other hand, the fluctuation $\mathbf{d}^1$ is $(-\sqrt{2}, -\sqrt{2}, \sqrt{2}, 0)$, which has energy

$$\mathcal{E}_{\mathbf{d}^1} = 2 + 2 + 2 + 0 = 6.$$

Thus the energy of the trend $\mathbf{a}^1$ accounts for $440/446 = 98.7\%$ of the total energy of the signal. In other words, the 1-level Haar transform has redistributed the energy of $\mathbf{f}$ so that over 98% is concentrated into the subsignal $\mathbf{a}^1$ which is half the length of $\mathbf{f}$. For obvious reasons, this is called *compaction of energy.* As another example, consider the signal $\mathbf{f}$ graphed in Figure 1.1(a) and its 1-level Haar transform shown in Figure 1.1(b). In this case, we find that the energy of the signal $\mathbf{f}$ is 127.308 while the energy of its first trend $\mathbf{a}^1$ is 127.305. Thus 99.998% of the total energy is compacted into the half-length subsignal $\mathbf{a}^1$. By examining the graph in Figure 1.1(b) it is easy to see why such a phenomenal energy compaction has occurred; the values of the fluctuation $\mathbf{d}^1$ are so small, relative to the much larger values of the trend $\mathbf{a}^1$, that its energy $\mathcal{E}_{\mathbf{d}^1}$ contributes only a small fraction of the total energy $\mathcal{E}_{\mathbf{a}^1} + \mathcal{E}_{\mathbf{d}^1}$.

These two examples illustrate the following general principle:

Compaction of Energy. *The energy of the trend subsignal* $\mathbf{a}^1$ *accounts for a large percentage of the energy of the transformed signal* $(\mathbf{a}^1 \mid \mathbf{d}^1)$.

Compaction of Energy will occur whenever the magnitudes of the fluctuation's values are significantly smaller than the trend's values (recall the Small Fluctuations Feature from the last section).

In Section 1.5, we shall describe how compaction of energy provides a framework for applying the Haar transform to compress signals. We now turn to a discussion of how the Haar transform can be extended to multiple levels, thereby increasing the energy compaction of signals.

Haar transform, multiple levels

Once we have performed a 1-level Haar transform, then it is easy to repeat the process and perform multiple level Haar transforms. After performing a 1-level Haar transform of a signal $\mathbf{f}$ we obtain a first trend $\mathbf{a}^1$ and a first fluctuation $\mathbf{d}^1$. The second level of a Haar transform is then performed by computing a second trend $\mathbf{a}^2$ and a second fluctuation $\mathbf{d}^2$ *for the first trend* $\mathbf{a}^1$ *only.*

For example, if $\mathbf{f} = (4, 6, 10, 12, 8, 6, 5, 5)$ is the signal considered above, then we found that its first trend is $\mathbf{a}^1 = (5\sqrt{2}, 11\sqrt{2}, 7\sqrt{2}, 5\sqrt{2})$. To get the second trend $\mathbf{a}^2$ we apply Formula (1.2) *to the values of* $\mathbf{a}^1$. That is, we add successive pairs of values of $\mathbf{a}^1$ and divide by $\sqrt{2}$ as indicated in the following diagram:

$$\begin{array}{lccccccc} \mathbf{a}^1: & 5\sqrt{2} & & 11\sqrt{2} & 7\sqrt{2} & & 5\sqrt{2} \\ & & \searrow\swarrow & & & \searrow\swarrow & \\ \mathbf{a}^2: & & 16 & & & 12 & \end{array}$$

And to get the second fluctuation $\mathbf{d}^2$ we subtract successive pairs of values of $\mathbf{a}^1$ and divide by $\sqrt{2}$ as indicated in this diagram:

$$\begin{array}{lccccccc} \mathbf{a}^1: & 5\sqrt{2} & & 11\sqrt{2} & 7\sqrt{2} & & 5\sqrt{2} \\ & & \searrow\swarrow & & & \searrow\swarrow & \\ \mathbf{d}^2: & & -6 & & & 2 & \end{array}$$

Thus the 2-level Haar transform of $\mathbf{f}$ is the signal

$$(\mathbf{a}^2 \,|\, \mathbf{d}^2 \,|\, \mathbf{d}^1) = (16, 12 \,|\, -6, 2 \,|\, -\sqrt{2}, -\sqrt{2}, \sqrt{2}, 0).$$

For this signal $\mathbf{f}$, a 3-level Haar transform can also be done, and the result is

$$(\mathbf{a}^3 \,|\, \mathbf{d}^3 \,|\, \mathbf{d}^2 \,|\, \mathbf{d}^1) = (14\sqrt{2} \,|\, 2\sqrt{2} \,|\, -6, 2 \,|\, -\sqrt{2}, -\sqrt{2}, \sqrt{2}, 0).$$

It is interesting to calculate the energy compaction that has occurred with the 2-level and 3-level Haar transforms that we just computed. First, we know that $\mathcal{E}_{(\mathbf{a}^2 \,|\, \mathbf{d}^2 \,|\, \mathbf{d}^1)} = 446$ because of Conservation of Energy. Second, we compute that $\mathcal{E}_{\mathbf{a}^2} = 400$. Thus the 2-level Haar transformed signal $(\mathbf{a}^2 \,|\, \mathbf{d}^2 \,|\, \mathbf{d}^1)$ has almost 90% of the total energy of $\mathbf{f}$ contained in the second trend $\mathbf{a}^2$ which is 1/4 of the length of $\mathbf{f}$. This is a further compaction, or *localization,* of the energy of $\mathbf{f}$. Furthermore, $\mathcal{E}_{\mathbf{a}^3} = 392$; thus $\mathbf{a}^3$ contains 87.89% of the total energy of $\mathbf{f}$. This is even further compaction; the 3-level Haar transform $(\mathbf{a}^3 \,|\, \mathbf{d}^3 \,|\, \mathbf{d}^2 \,|\, \mathbf{d}^1)$ has almost 88% of the total energy of $\mathbf{f}$ contained in the third trend $\mathbf{a}^3$ which is 1/8 the length of $\mathbf{f}$.

For those readers who are familiar with Quantum Theory, there is an interesting phenomenon here that is worth noting. By Heisenberg's Uncertainty Principle, it is impossible to localize a fixed amount of energy into an arbitrarily small time interval. This provides an explanation for why the energy percentage dropped from 98% to 90% when the second-level Haar transform was computed, and from 90% to 88% when the third-level Haar transform was computed. When we attempt to squeeze the energy into ever smaller time intervals, it is inevitable that some energy leaks out.

As another example of how the Haar transform redistributes and localizes the energy in a signal, consider the graphs shown in Figure 1.2. In Figure 1.2(a) we show a signal, and in Figure 1.2(b) we show the 2-level Haar transform of this signal. In Figures 1.2(c) and (d) we show the respective cumulative energy profiles of these two signals. By the *cumulative energy profile* of a signal $\mathbf{f}$ we mean the signal defined by

$$\left(\frac{f_1^2}{\mathcal{E}_\mathbf{f}}, \frac{f_1^2+f_2^2}{\mathcal{E}_\mathbf{f}}, \frac{f_1^2+f_2^2+f_3^2}{\mathcal{E}_\mathbf{f}}, \dots, 1\right).$$

The cumulative energy profile of $\mathbf{f}$ thus provides a summary of the accumulation of energy in the signal as time proceeds. As can be seen from comparing the two profiles in Figure 1.2, the 2-level Haar transform has redistributed and localized the energy of the original signal.

Justification of Energy Conservation

We close this section with a brief justification of the Conservation of Energy property of the Haar transform. First, we observe that the terms a_1^2 and d_1^2 in the formula $\mathcal{E}_{(\mathbf{a}^1\,|\,\mathbf{d}^1)} = a_1^2 + \cdots + a_{N/2}^2 + d_1^2 + \cdots + d_{N/2}^2$ add up as follows:

$$\begin{aligned} a_1^2 + d_1^2 &= \left[\frac{f_1+f_2}{\sqrt{2}}\right]^2 + \left[\frac{f_1-f_2}{\sqrt{2}}\right]^2 \\ &= \frac{f_1^2+2f_1f_2+f_2^2}{2} + \frac{f_1^2-2f_1f_2+f_2^2}{2} \\ &= f_1^2+f_2^2. \end{aligned}$$

Similarly, $a_m^2 + d_m^2 = f_{2m-1}^2 + f_{2m}^2$ for all other values of m. Therefore, by adding a_m^2 and d_m^2 successively for each m, we find that

$$a_1^2 + \cdots + a_{N/2}^2 + d_1^2 + \cdots + d_{N/2}^2 = f_1^2 + \cdots + f_N^2.$$

In other words, $\mathcal{E}_{(\mathbf{a}^1\,|\,\mathbf{d}^1)} = \mathcal{E}_\mathbf{f}$, which justifies the Conservation of Energy property.

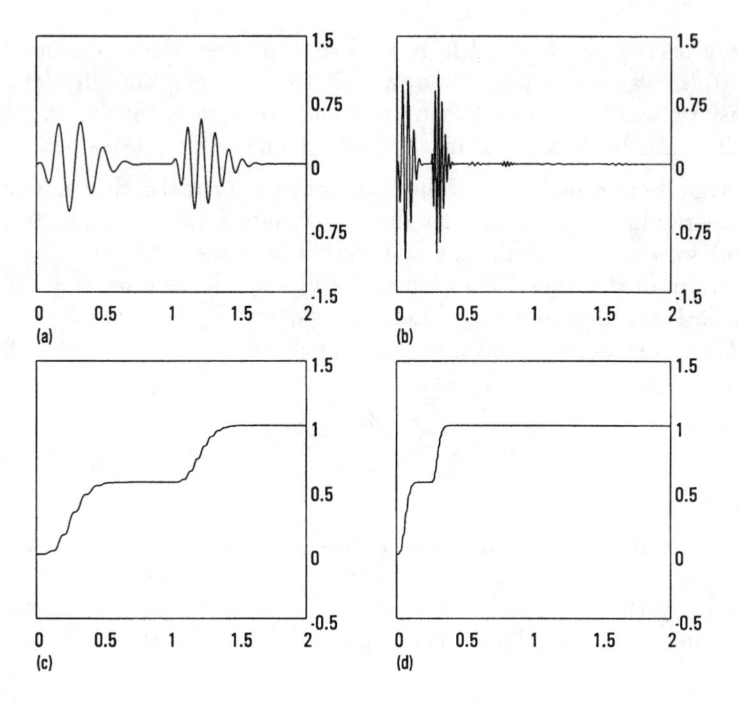

FIGURE 1.2
(a) Signal. (b) 2-level Haar transform of signal. (c) Cumulative energy profile of Signal. (d) Cumulative energy profile of 2-level Haar transform.

1.3 Haar wavelets

In this section we discuss the simplest wavelets, the Haar wavelets. This material will set the stage for the more sophisticated Daubechies wavelets described in the next chapter.

We begin by discussing the 1-level *Haar wavelets.* These wavelets are defined as

$$\begin{aligned}
\mathbf{W}_1^1 &= \left(\frac{1}{\sqrt{2}}, \frac{-1}{\sqrt{2}}, 0, 0, \ldots, 0\right) \\
\mathbf{W}_2^1 &= \left(0, 0, \frac{1}{\sqrt{2}}, \frac{-1}{\sqrt{2}}, 0, 0, \ldots, 0\right) \\
&\vdots \\
\mathbf{W}_{N/2}^1 &= \left(0, 0, \ldots, 0, \frac{1}{\sqrt{2}}, \frac{-1}{\sqrt{2}}\right).
\end{aligned} \tag{1.8}$$

These 1-level Haar wavelets have a number of interesting properties. First, they each have an energy of 1. Second, they each consist of a rapid fluctuation between just two non-zero values, $\pm 1/\sqrt{2}$, with an average value of 0. Hence the name *wavelet*. Finally, they all are very similar to each other in that they are each a translation in time by an even number of time-units of the first Haar wavelet $\mathbf{W}_1^1$. The second Haar wavelet $\mathbf{W}_2^1$ is a translation forward in time by two units of $\mathbf{W}_1^1$, and $\mathbf{W}_3^1$ is a translation forward in time by four units of $\mathbf{W}_1^1$, and so on.

The reason for introducing the 1-level Haar wavelets is that we can express the 1-level fluctuation subsignal in a simpler form by making use of scalar products with these wavelets. The scalar product is a fundamental operation on two signals, and is defined as follows.

Scalar product: *The scalar product* $\mathbf{f} \cdot \mathbf{g}$ *of the signals* $\mathbf{f} = (f_1, f_2, \ldots, f_N)$ *and* $\mathbf{g} = (g_1, g_2, \ldots, g_N)$ *is defined by*

$$\mathbf{f} \cdot \mathbf{g} = f_1 g_1 + f_2 g_2 + \cdots + f_N g_N. \tag{1.9}$$

Using the 1-level Haar wavelets, we can express the values for the first fluctuation subsignal $\mathbf{d}^1$ as scalar products. For example,

$$d_1 = \frac{f_1 - f_2}{\sqrt{2}} = \mathbf{f} \cdot \mathbf{W}_1^1.$$

Similarly, $d_2 = \mathbf{f} \cdot \mathbf{W}_2^1$, and so on. We can summarize Formula (1.3) in terms of scalar products with the 1-level Haar wavelets:

$$d_m = \mathbf{f} \cdot \mathbf{W}_m^1 \tag{1.10}$$

for $m = 1, 2, \ldots, N/2$.

We can also use the idea of scalar products to restate the Small Fluctuations Feature from Section 1.1 in a more precise form. If we say that the *support* of each Haar wavelet is the set of two time-indices where the wavelet is non-zero, then we have the following more precise version of the Small Fluctuations Feature:

Property 1. *If a signal* $\mathbf{f}$ *is (approximately) constant over the support of a 1-level Haar wavelet* $\mathbf{W}_k^1$*, then the fluctuation value* $d_k = \mathbf{f} \cdot \mathbf{W}_k^1$ *is (approximately) zero.*

This property will be considerably strengthened in the next chapter.

Note: *From now on, we shall refer to the set of time-indices* m *where* $f_m \neq 0$ *as the* support *of a signal* $\mathbf{f}$.

We can also express the 1-level trend values as scalar products with certain elementary signals. These elementary signals are called 1-level *Haar scaling signals,* and they are defined as

$$\mathbf{V}_1^1 = \left(\frac{1}{\sqrt{2}}, \frac{1}{\sqrt{2}}, 0, 0, \dots, 0\right)$$
$$\mathbf{V}_2^1 = \left(0, 0, \frac{1}{\sqrt{2}}, \frac{1}{\sqrt{2}}, 0, 0, \dots, 0\right)$$
$$\vdots$$
$$\mathbf{V}_{N/2}^1 = \left(0, 0, \dots, 0, \frac{1}{\sqrt{2}}, \frac{1}{\sqrt{2}}\right). \tag{1.11}$$

Using these Haar scaling signals, the values $a_1, \dots, a_{N/2}$ for the first trend are expressed as scalar products:

$$a_m = \mathbf{f} \cdot \mathbf{V}_m^1 \tag{1.12}$$

for $m = 1, 2, \dots, N/2$.

The Haar scaling signals are quite similar to the Haar wavelets. They all have energy 1 and have a support consisting of just two consecutive time-indices. In fact, they are all translates by an even multiple of time-units of the first scaling signal $\mathbf{V}_1^1$. Unlike the Haar wavelets, however, the average values of the Haar scaling signals are not zero. In fact, they each have an average value of $1/\sqrt{2}$.

The ideas discussed above extend to every level. For simplicity, we restrict our discussion to the second level. The 2-level Haar scaling signals are defined by

$$\mathbf{V}_1^2 = \left(\frac{1}{2}, \frac{1}{2}, \frac{1}{2}, \frac{1}{2}, 0, 0 \dots, 0\right)$$
$$\mathbf{V}_2^2 = \left(0, 0, 0, 0, \frac{1}{2}, \frac{1}{2}, \frac{1}{2}, \frac{1}{2}, 0, 0, \dots, 0\right)$$
$$\vdots$$
$$\mathbf{V}_{N/4}^2 = \left(0, 0, \dots, 0, \frac{1}{2}, \frac{1}{2}, \frac{1}{2}, \frac{1}{2}\right). \tag{1.13}$$

These scaling signals are all translations by multiples of four time-units of the first scaling signal $\mathbf{V}_1^2$, and they all have energy 1 and average value 1/2. Furthermore, the values of the 2-level trend $\mathbf{a}^2$ are scalar products of these scaling signals with the signal $\mathbf{f}$. That is, $\mathbf{a}^2$ satisfies

$$\mathbf{a}^2 = \left(\mathbf{f} \cdot \mathbf{V}_1^2, \mathbf{f} \cdot \mathbf{V}_2^2, \dots, \mathbf{f} \cdot \mathbf{V}_{N/4}^2\right). \tag{1.14}$$

Likewise, the 2-level Haar wavelets are defined by

$$\mathbf{W}_1^2 = \left(\frac{1}{2}, \frac{1}{2}, \frac{-1}{2}, \frac{-1}{2}, 0, 0 \dots, 0\right)$$

$$\mathbf{W}_2^2 = \left(0, 0, 0, 0, \frac{1}{2}, \frac{1}{2}, \frac{-1}{2}, \frac{-1}{2}, 0, 0, \ldots, 0\right)$$

$$\vdots$$

$$\mathbf{W}_{N/4}^2 = \left(0, 0, \ldots, 0, \frac{1}{2}, \frac{1}{2}, \frac{-1}{2}, \frac{-1}{2}\right). \tag{1.15}$$

These wavelets all have supports of length 4, since they are all translations by multiples of four time-units of the first wavelet $\mathbf{W}_1^2$. They also all have energy 1 and average value 0. Using scalar products, the 2-level fluctuation $\mathbf{d}^2$ satisfies

$$\mathbf{d}^2 = \left(\mathbf{f} \cdot \mathbf{W}_1^2, \mathbf{f} \cdot \mathbf{W}_2^2, \ldots, \mathbf{f} \cdot \mathbf{W}_{N/4}^2\right). \tag{1.16}$$

1.4 Multiresolution analysis

In the previous section we discussed how the Haar transform can be described using scalar products with scaling signals and wavelets. In this section we discuss how the inverse Haar transform can also be described in terms of these same elementary signals. This discussion will show how discrete signals are synthesized by beginning with a very low resolution signal and successively adding on details to create higher resolution versions, ending with a complete synthesis of the signal at the finest resolution. This is known as *multiresolution analysis* (MRA). MRA is the heart of wavelet analysis.

In order to make these ideas precise, we must first discuss some elementary operations that can be performed on signals. Given two signals $\mathbf{f} = (f_1, f_2, \ldots, f_N)$ and $\mathbf{g} = (g_1, g_2, \ldots, g_N)$, we can perform the following elementary algebraic operations:

Addition and Subtraction: The *sum* $\mathbf{f} + \mathbf{g}$ of the signals $\mathbf{f}$ and $\mathbf{g}$ is defined by adding their values:

$$\mathbf{f} + \mathbf{g} = (f_1 + g_1, f_2 + g_2, \ldots, f_N + g_N). \tag{1.17}$$

Their *difference* $\mathbf{f} - \mathbf{g}$ is defined by subtracting their values:

$$\mathbf{f} - \mathbf{g} = (f_1 - g_1, f_2 - g_2, \ldots, f_N - g_N). \tag{1.18}$$

Constant multiple: A signal $\mathbf{f}$ is multiplied by a constant c by multiplying each of its values by c. That is,

$$c\mathbf{f} = (cf_1, cf_2, \ldots, cf_N). \tag{1.19}$$

For example, by repeatedly applying the addition operation, we can express a signal $\mathbf{f} = (f_1, f_2, \ldots, f_N)$ as follows:

$$\mathbf{f} = (f_1, 0, 0, \ldots, 0) + (0, f_2, 0, 0, \ldots, 0) + \cdots + (0, 0, \ldots, 0, f_N).$$

Then, by applying the constant multiple operation to each of the signals on the right side of this last equation, we obtain

$$\mathbf{f} = f_1(1, 0, 0, \ldots, 0) + f_2(0, 1, 0, 0, \ldots, 0) + \cdots + f_N(0, 0, \ldots, 0, 1).$$

This formula is a very natural one; it amounts to expressing $\mathbf{f}$ as a sum of its individual values at each discrete instant of time.

If we define the elementary signals $\mathbf{V}_1^0, \mathbf{V}_2^0, \ldots, \mathbf{V}_N^0$ as

$$\begin{aligned}
\mathbf{V}_1^0 &= (1, 0, 0, \ldots, 0) \\
\mathbf{V}_2^0 &= (0, 1, 0, 0, \ldots, 0) \\
&\ \ \vdots \\
\mathbf{V}_N^0 &= (0, 0, \ldots, 0, 1)
\end{aligned} \tag{1.20}$$

then the last formula for $\mathbf{f}$ can be rewritten as

$$\mathbf{f} = f_1\mathbf{V}_1^0 + f_2\mathbf{V}_2^0 + \cdots + f_N\mathbf{V}_N^0. \tag{1.21}$$

Formula (1.21) is called the *natural expansion* of a signal $\mathbf{f}$ in terms of the *natural basis* of signals $\mathbf{V}_1^0, \mathbf{V}_2^0, \ldots, \mathbf{V}_N^0$. We shall now show that the Haar MRA involves expressing $\mathbf{f}$ as a sum of constant multiples of a different basis set of elementary signals, the Haar wavelets and scaling signals defined in the previous section.

In the previous section, we showed how to express the 1-level Haar transform in terms of wavelets and scaling signals. It is also possible to express the inverse of the 1-level Haar transform in terms of these same elementary signals. This leads to the first level of the Haar MRA. To define this first level Haar MRA we make use of (1.6) to express a signal $\mathbf{f}$ as

$$\begin{aligned}
\mathbf{f} = &\left(\frac{a_1}{\sqrt{2}}, \frac{a_1}{\sqrt{2}}, \frac{a_2}{\sqrt{2}}, \frac{a_2}{\sqrt{2}}, \ldots, \frac{a_{N/2}}{\sqrt{2}}, \frac{a_{N/2}}{\sqrt{2}}\right) \\
&+ \left(\frac{d_1}{\sqrt{2}}, \frac{-d_1}{\sqrt{2}}, \frac{d_2}{\sqrt{2}}, \frac{-d_2}{\sqrt{2}}, \ldots, \frac{d_{N/2}}{\sqrt{2}}, \frac{-d_{N/2}}{\sqrt{2}}\right).
\end{aligned}$$

This formula shows that the signal $\mathbf{f}$ can be expressed as the sum of two signals that we shall call the first averaged signal and the first detail signal. That is, we have

$$\mathbf{f} = \mathbf{A}^1 + \mathbf{D}^1 \tag{1.22}$$

where the signal $\mathbf{A}^1$ is called the *first averaged signal* and is defined by

$$\mathbf{A}^1 = \left(\frac{a_1}{\sqrt{2}}, \frac{a_1}{\sqrt{2}}, \frac{a_2}{\sqrt{2}}, \frac{a_2}{\sqrt{2}}, \dots, \frac{a_{N/2}}{\sqrt{2}}, \frac{a_{N/2}}{\sqrt{2}}\right) \tag{1.23}$$

and the signal $\mathbf{D}^1$ is called the *first detail signal* and is defined by

$$\mathbf{D}^1 = \left(\frac{d_1}{\sqrt{2}}, \frac{-d_1}{\sqrt{2}}, \frac{d_2}{\sqrt{2}}, \frac{-d_2}{\sqrt{2}}, \dots, \frac{d_{N/2}}{\sqrt{2}}, \frac{-d_{N/2}}{\sqrt{2}}\right). \tag{1.24}$$

Using Haar scaling signals and wavelets, and using the basic elementary algebraic operations with signals, the averaged and detail signals are expressible as

$$\mathbf{A}^1 = a_1\mathbf{V}_1^1 + a_2\mathbf{V}_2^1 + \cdots + a_{N/2}\mathbf{V}_{N/2}^1, \tag{1.25a}$$

$$\mathbf{D}^1 = d_1\mathbf{W}_1^1 + d_2\mathbf{W}_2^1 + \cdots + d_{N/2}\mathbf{W}_{N/2}^1. \tag{1.25b}$$

Applying the scalar product formulas for the coefficients in Equations (1.10) and (1.12), we can rewrite these last two formulas as follows

$$\mathbf{A}^1 = (\mathbf{f}\cdot\mathbf{V}_1^1)\mathbf{V}_1^1 + (\mathbf{f}\cdot\mathbf{V}_2^1)\mathbf{V}_2^1 + \cdots + (\mathbf{f}\cdot\mathbf{V}_{N/2}^1)\mathbf{V}_{N/2}^1,$$

$$\mathbf{D}^1 = (\mathbf{f}\cdot\mathbf{W}_1^1)\mathbf{W}_1^1 + (\mathbf{f}\cdot\mathbf{W}_2^1)\mathbf{W}_2^1 + \cdots + (\mathbf{f}\cdot\mathbf{W}_{N/2}^1)\mathbf{W}_{N/2}^1.$$

These formulas show that the averaged signal is a combination of Haar scaling signals, with the values of the first trend subsignal as coefficients; and that the detail signal is a combination of Haar wavelets, with the values of the first fluctuation subsignal as coefficients.

As an example of these ideas, consider the signal

$$\mathbf{f} = (4, 6, 10, 12, 8, 6, 5, 5).$$

In Section 1.1 we found that its first trend subsignal was

$$\mathbf{a}^1 = (5\sqrt{2}, 11\sqrt{2}, 7\sqrt{2}, 5\sqrt{2}).$$

Applying Formula (1.23), the averaged signal is

$$\mathbf{A}^1 = (5, 5, 11, 11, 7, 7, 5, 5). \tag{1.26}$$

Notice how the first averaged signal consists of the repeated average values 5, 5, and 11, 11, and 7, 7, and 5, 5 about which the values of $\mathbf{f}$ fluctuate. Using Formula (1.25a), the first averaged signal can also be expressed in terms of Haar scaling signals as

$$\mathbf{A}^1 = 5\sqrt{2}\,\mathbf{V}_1^1 + 11\sqrt{2}\,\mathbf{V}_2^1 + 7\sqrt{2}\,\mathbf{V}_3^1 + 5\sqrt{2}\,\mathbf{V}_4^1.$$

Comparing these last two equations we can see that *the positions of the repeated averages correspond precisely with the supports of the scaling signals.*

We also found in Section 1.1 that the first fluctuation signal for $\mathbf{f}$ was $\mathbf{d}^1 = (-\sqrt{2}, -\sqrt{2}, \sqrt{2}, 0)$. Formula (1.24) then yields

$$\mathbf{D}^1 = (-1, 1, -1, 1, 1, -1, 0, 0).$$

Thus, using the result for $\mathbf{A}^1$ computed above, we have

$$\mathbf{f} = (5, 5, 11, 11, 7, 7, 5, 5) + (-1, 1, -1, 1, 1, -1, 0, 0).$$

This equation illustrates the basic idea of MRA. The signal $\mathbf{f}$ is expressed as a sum of a lower resolution, or averaged, signal $(5, 5, 11, 11, 7, 7, 5, 5)$ added with a signal $(-1, 1, -1, 1, 1, -1, 0, 0)$ made up of fluctuations or details. These fluctuations provide the added details necessary to produce the full resolution signal $\mathbf{f}$.

For this example, using Formula (1.25b), the first detail signal can also be expressed in terms of Haar wavelets as

$$\mathbf{D}^1 = -\sqrt{2}\,\mathbf{W}_1^1 - \sqrt{2}\,\mathbf{W}_2^1 + \sqrt{2}\,\mathbf{W}_3^1 + 0\,\mathbf{W}_4^1.$$

This formula shows that the values of $\mathbf{D}^1$ occur in successive pairs of rapidly fluctuating values positioned at the supports of the Haar wavelets.

Multiresolution analysis, multiple levels

In the discussion above, we described the first level of the Haar MRA of a signal. This idea can be extended to further levels, as many levels as the number of times that the signal length can be divided by 2.

The second level of a MRA of a signal $\mathbf{f}$ involves expressing $\mathbf{f}$ as

$$\mathbf{f} = \mathbf{A}^2 + \mathbf{D}^2 + \mathbf{D}^1. \tag{1.27}$$

Here $\mathbf{A}^2$ is the second averaged signal and $\mathbf{D}^2$ is the second detail signal. Comparing Formulas (1.22) and (1.27) we see that

$$\mathbf{A}^1 = \mathbf{A}^2 + \mathbf{D}^2. \tag{1.28}$$

This formula expresses the fact that computing the second averaged signal $\mathbf{A}^2$ and second detail signal $\mathbf{D}^2$ simply consists of performing a first level MRA of the signal $\mathbf{A}^1$. Because of this, it follows that the second level averaged signal $\mathbf{A}^2$ satisfies

$$\mathbf{A}^2 = (\mathbf{f} \cdot \mathbf{V}_1^2)\mathbf{V}_1^2 + (\mathbf{f} \cdot \mathbf{V}_2^2)\mathbf{V}_2^2 + \cdots + (\mathbf{f} \cdot \mathbf{V}_{N/4}^2)\mathbf{V}_{N/4}^2$$

and the second level detail signal $\mathbf{D}^2$ satisfies

$$\mathbf{D}^2 = (\mathbf{f} \cdot \mathbf{W}_1^2)\mathbf{W}_1^2 + (\mathbf{f} \cdot \mathbf{W}_2^2)\mathbf{W}_2^2 + \cdots + (\mathbf{f} \cdot \mathbf{W}_{N/4}^2)\mathbf{W}_{N/4}^2.$$

For example, if $\mathbf{f} = (4, 6, 10, 12, 8, 6, 5, 5)$, then we that $\mathbf{a}^2 = (16, 12)$. Therefore

$$\mathbf{A}^2 = 16\left(\frac{1}{2}, \frac{1}{2}, \frac{1}{2}, \frac{1}{2}, 0, 0, 0, 0\right) + 12\left(0, 0, 0, 0, \frac{1}{2}, \frac{1}{2}, \frac{1}{2}, \frac{1}{2}\right)$$
$$= (8, 8, 8, 8, 6, 6, 6, 6). \qquad (1.29)$$

It is interesting to compare the equations in (1.26) and (1.29). The second averaged signal $\mathbf{A}^2$ has values created from averages that involve twice as many values as the averages that created $\mathbf{A}^1$. Therefore, the second averaged signal reflects more long term trends than those reflected in the first averaged signal. Consequently, these averages are repeated for twice as many time-units.

We also found in Section 1.2 that this signal $\mathbf{f} = (4, 6, 10, 12, 8, 6, 5, 5)$ has the second fluctuation $\mathbf{d}^2 = (-6, 2)$. Consequently

$$\mathbf{D}^2 = -6\left(\frac{1}{2}, \frac{1}{2}, \frac{-1}{2}, \frac{-1}{2}, 0, 0, 0, 0\right) + 2\left(0, 0, 0, 0, \frac{1}{2}, \frac{1}{2}, \frac{-1}{2}, \frac{-1}{2}\right)$$
$$= (-3, -3, 3, 3, 1, 1, -1, -1).$$

We found above that $\mathbf{D}^1 = (-1, 1, -1, 1, 1, -1, 0, 0)$. Hence

$$\begin{aligned}\mathbf{f} &= \mathbf{A}^2 + \mathbf{D}^2 + \mathbf{D}^1 \\ &= (8, 8, 8, 8, 6, 6, 6, 6) + (-3, -3, 3, 3, 1, 1, -1, -1) \\ &\qquad + (-1, 1, -1, 1, 1, -1, 0, 0).\end{aligned}$$

This formula further illustrates the idea of MRA. The full resolution signal $\mathbf{f}$ is produced from a very low resolution, averaged signal $\mathbf{A}^2$ consisting of repetitions of the two averaged values, 8 and 6, to which are added two detail signals. The first addition supplements this averaged signal with enough details to produce the next higher resolution averaged signal $(5, 5, 11, 11, 7, 7, 5, 5)$, and the second addition then supplies enough further details to produce the full resolution signal $\mathbf{f}$.

In general, if the number N of signal values is divisible k times by 2, then a k-level MRA:

$$\mathbf{f} = \mathbf{A}^k + \mathbf{D}^k + \cdots + \mathbf{D}^2 + \mathbf{D}^1$$

can be performed on the signal $\mathbf{f}$. Rather than subjecting the reader to the gory details, we conclude by describing a computer example generated using FAWAV. In Figure 1.3 we show a 10-level Haar MRA of the signal $\mathbf{f}$ shown in Figure 1.1(a). This signal has 2^{10} values so 10 levels of MRA are possible. On the top of Figure 1.3(a), the graph of $\mathbf{A}^{10}$ is shown; it consists of a single value repeated 2^{10} times. This value is the average of

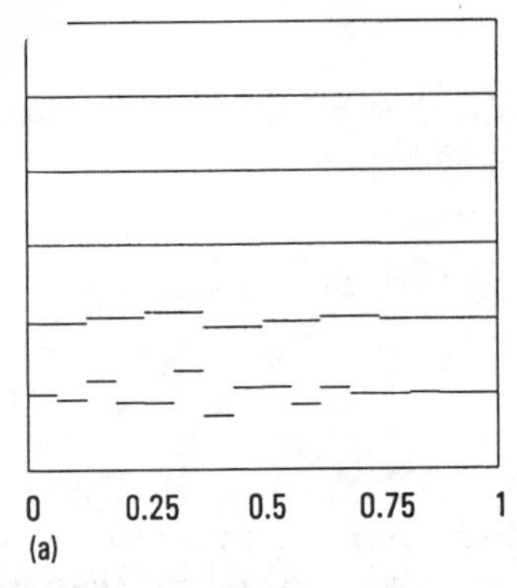

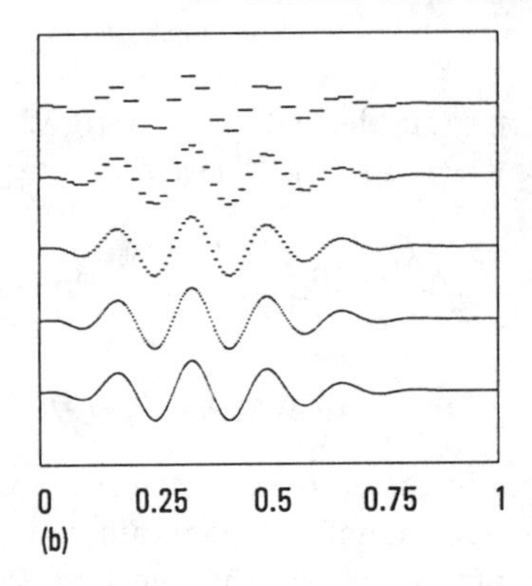

FIGURE 1.3
Haar MRA of the signal in Figure 1.1(a). The graphs are of the ten averaged signals $\mathbf{A}^{10}$ through $\mathbf{A}^1$. Beginning with the signal $\mathbf{A}^{10}$ on the top left down to $\mathbf{A}^6$ on the bottom left, then $\mathbf{A}^5$ on the top right down to $\mathbf{A}^1$ on the bottom right.

all 2^{10} values of the signal $\mathbf{f}$. The graph directly below it is of the signal $\mathbf{A}^9$ which equals $\mathbf{A}^{10}$ plus the details in $\mathbf{D}^{10}$. Each successive averaged signal is shown, from $\mathbf{A}^{10}$ through $\mathbf{A}^1$. By successively adding on details, the full signal in Figure 1.1(a) is systematically constructed in all its complexity.

1.5 Compression of audio signals

In Section 1.2 we saw that the Haar transform can be used to localize the energy of a signal into a shorter subsignal. In this section we show how this redistribution of energy can be used to compress audio signals. By compressing an audio signal we mean converting the signal data into a new format that requires less bits to transmit. When we use the term, *audio signal,* we are speaking somewhat loosely. Many of the signals we have in mind are indeed the result of taking discrete samples of a sound signal—as in the data in a computer audio file, or on a compact disc—but the techniques developed here also apply to digital data transmissions and to other digital signals, such as digitized electrocardiograms or digitized electroencephalograms.

There are two basic categories of compression techniques. The first category is *lossless compression.* Lossless compression methods achieve completely error free decompression of the original signal. Typical lossless methods are Huffman compression, LZW compression, arithmetic compression, or run-length compression. Combinations of these techniques are used in popular lossless compression programs, such as the kind that produce `.zip` files. Unfortunately, the compression ratios that can be obtained with lossless methods are rarely more than 2:1 for audio files consisting of music or speech.

The second category is *lossy compression.* A lossy compression method is one which produces inaccuracies in the decompressed signal. Lossy techniques are used when these inaccuracies are so small as to be imperceptible. The advantage of lossy techniques over lossless ones is that much higher compression ratios can be attained. With wavelet compression methods, which are lossy, if we are willing to accept the slight inaccuracies in the decompressed signal, then we can obtain compression ratios of 10:1, or 20:1, or as high as 50:1 or even 100:1.

In order to illustrate the general principles of wavelet compression of signals, we shall examine, in a somewhat simplified way, how the Haar wavelet transform can be used to compress some test signals. For example, Signal 1 in Figure 1.4(a) can be very effectively compressed using the Haar transform. Although Signal 1 is not a very representative audio signal, it is representative of a portion of a digital data transmission. This signal has 1024 values equally spaced over the time interval $[0, 20)$. Most of these values are constant over long stretches, and that is the principal reason that Signal 1 can be compressed effectively with the Haar transform. Signal 2 in Figure 1.5(a), however, will not compress nearly so well; this signal requires the more sophisticated wavelet transforms described in the next chapter.

The basic steps for wavelet compression are as follows:

Method of Wavelet Transform Compression

Step 1. Perform a wavelet transform of the signal.

Step 2. Set equal to 0 all values of the wavelet transform which are insignificant, i.e., which lie below some *threshold value.*

Step 3. Transmit only the significant, non-zero values of the transform obtained from Step 2. This should be a much smaller data set than the original signal.

Step 4. At the receiving end, perform the inverse wavelet transform of the data transmitted in Step 3, assigning zero values to the insignificant values which were not transmitted. This decompression step produces an approximation of the original signal.

In this chapter we shall illustrate this method using the Haar wavelet transform. This initial discussion will be significantly deepened and generalized in the next chapter when we discuss this method of compression in terms of various Daubechies wavelet transforms.

Let's now examine a Haar wavelet transform compression of Signal 1. We begin with Step 1. Since Signal 1 consists of $1024 = 2^{10}$ values, we

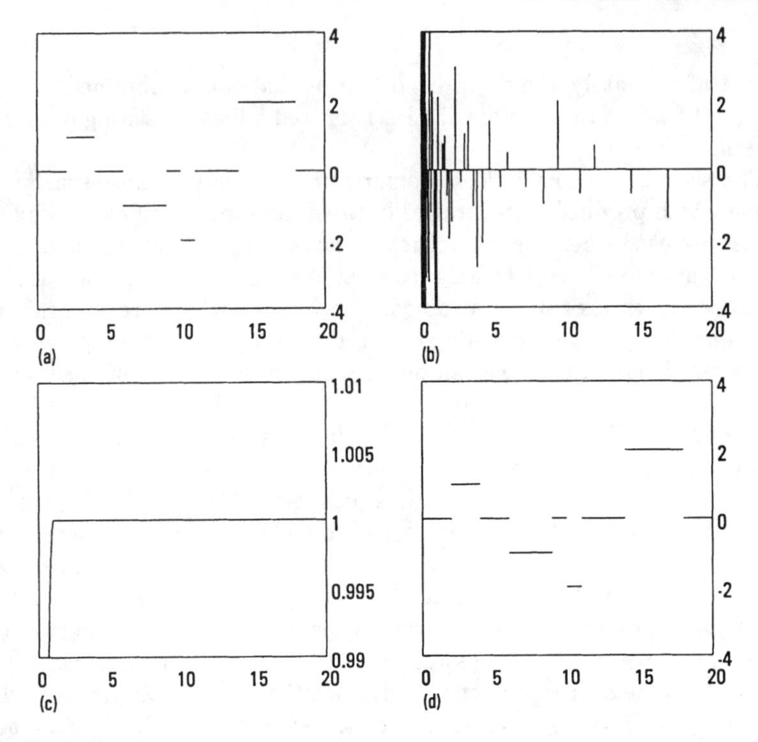

FIGURE 1.4
(a) Signal 1, (b) 10-level Haar transform of Signal 1, (c) energy map of Haar transform, (d) 20:1 compression of Signal 1, 100% of energy.

can perform 10 levels of the Haar transform. This 10-level Haar transform is shown in Figure 1.4(b). Notice how a large portion of the Haar transform's values are 0, or very near 0, in magnitude. This fact provides the fundamental basis for performing an effective compression.

In order to choose a threshold value for Step 2, we proceed as follows. First, we arrange the magnitudes of the values of the Haar transform so that they are in decreasing order:

$$L_1 \geq L_2 \geq L_3 \geq \ldots \geq L_N$$

where L_1 is the largest absolute value of the Haar transform, L_2 is the next largest, etc. (In the event of a tie, we just leave those magnitudes in their original order.) We then compute the cumulative energy profile of this new signal:

$$\left(\frac{L_1^2}{\mathcal{E}_{\mathbf{f}}}, \frac{L_1^2+L_2^2}{\mathcal{E}_{\mathbf{f}}}, \frac{L_1^2+L_2^2+L_3^2}{\mathcal{E}_{\mathbf{f}}}, \ldots, 1\right).$$

For Signal 1, we show a graph of this energy profile—which we refer to as the *energy map* of the Haar transform—in Figure 1.4(c). Notice that the energy map very quickly reaches its maximum value of 1. In fact, using FAWAV we find that

$$\frac{L_1^2 + L_2^2 + \ldots + L_{51}^2}{\mathcal{E}_{\mathbf{f}}} = .999996.$$

Consequently, if we choose a threshold T that is less than $L_{51} = .3536$, then the values of the transform that survive this threshold will account for essentially 100% of the energy of Signal 1.

We now turn to Step 3. In order to perform Step 3—transmitting only the significant transform values—an additional amount of information must be sent which indicates the positions of these significant transform values in the thresholded transform. This information is called the *significance map*. The values of this significance map are either 1 or 0: a value of 1 if the corresponding transform value survived the thresholding, a value of 0 if it did not. The significance map is therefore a string of N bits, where N is the length of the signal. For the case of Signal 1, with a threshold of .35, there are only 51 non-zero bits in the significance map out of a total of 1024 bits. Therefore, since most of this significance map consists of long stretches of zeros, it can be very effectively compressed using one of the lossless compression algorithms mentioned above. This compressed string of bits is then transmitted along with the non-zero values of the thresholded transform.

Finally, we arrive at Step 4. At the receiving end, the significance map is used to insert zeros in their proper locations in between the non-zero values in the thresholded transform, and then an inverse transform is computed to produce an approximation of the signal. For Signal 1 we show the approximation that results from using a threshold of .35 in Figure 1.4(d). This approximation used only 51 transform values; so it represents a compression of Signal 1 by a factor of 1024:51, i.e., a compression factor of 20:1. Since the compressed signal contains nearly 100% of the energy of the original signal, it is a very good approximation. In fact, the maximum error over all values is no more than 3.91×10^{-3}.

Life would be simpler if the Haar transform could be used so effectively for all signals. Unfortunately, if we try to use the Haar transform for threshold compression of Signal 2 in Figure 1.5(a), we get poor results. This signal, when played over a computer sound system, produces a sound similar to two low notes played on a clarinet. It has $4096 = 2^{12}$ values; so we can perform 12 levels of the Haar transform. In Figure 1.5(b) we show a plot of the 12-level Haar transform of Signal 2. It is clear from this plot that a large fraction of the Haar transform values have significant magnitude, significant enough that they are visible in the graph. In fact, the energy map for the transform of Signal 2, shown in Figure 1.5(c), exhibits a much slower increase towards 1 in comparison with the energy map for the transform of Signal 1. Therefore, many more transform values are needed

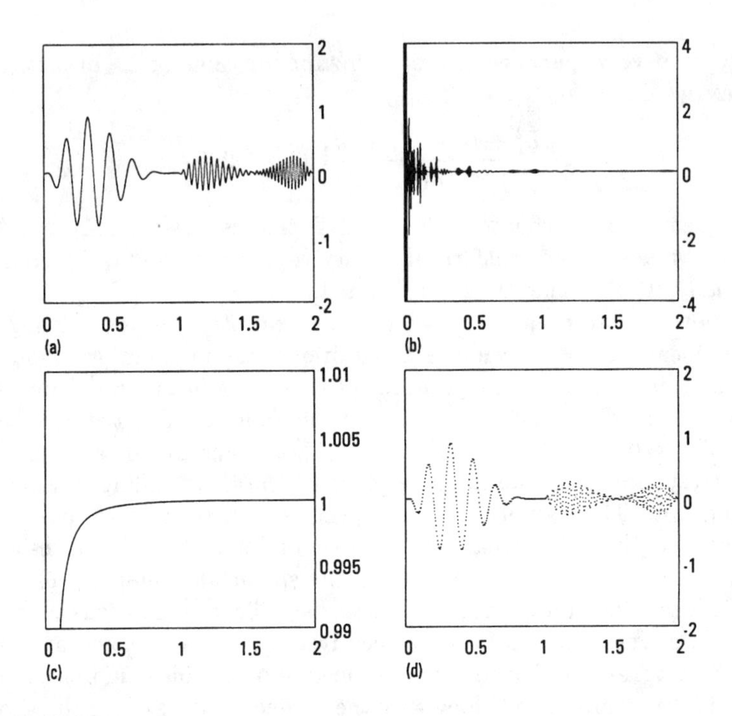

FIGURE 1.5
(a) Signal 2, (b) 12-level Haar transform of Signal 2, (c) energy map of Haar transform, (d) 10:1 compression of Signal 2, 99.6% of energy of Signal 2.

in order to capture a high percentage of the energy of Signal 2. In Figure 1.5(d), we show a 10:1 compression of Signal 2 which captures 99.6% of the energy of Signal 2. Comparing this compression with the original signal we see that it is a fairly poor approximation. Many of the signal values are clumped together in the compressed signal, producing a very ragged or jumpy approximation of the original signal. When this compressed version is played on a computer sound system, it produces a screechy "metallic" version of the two clarinet notes, which is not a very satisfying result. As a rule of thumb, we must capture at least 99.99% of the energy of the signal in order to produce an acceptable approximation, i.e., an approximation that is not perceptually different from the original. Achieving this accurate an approximation for Signal 2 requires at least 1782 transform values. Because Signal 2 itself has 4096 values, this is a compression ratio of only about 2.3:1, which is not very high. We shall see in the next chapter that Signal 2 can be compressed very effectively, but we shall need more high powered wavelet transforms to do it.

A note on quantization

The most serious oversimplification that we made in the discussion above is that we ignored the issue known as *quantization.* The term quantization is used whenever it is necessary to take into account the finite precision of numerical data handled by digital methods. For example, the numerical data used to generate the graphs of Signals 1 and 2 above were IEEE double precision numbers that use 8 bytes = 64 bits for each number. In order to compress this data even further, we can represent the wavelet transform coefficients using less bits. We shall address this issue of quantization in the next chapter when we look again at the problem of compression.

1.6 Removing noise from audio signals

In this section we shall begin our treatment of one of the most important aspects of signal processing, the removal of noise from signals. Our discussion in this section will introduce the fundamental ideas involved in the context of the Haar transform. In the next chapter we shall considerably deepen and generalize these ideas, in the context of the more powerful Daubechies wavelet transforms.

When a signal is received after transmission over some distance, it is frequently contaminated by noise. The term *noise* refers to any undesired change that has altered the values of the original signal. The simplest model for acquisition of noise by a signal is *additive* noise, which has the form

$$(\textit{contaminated signal}) = (\textit{original signal}) + (\textit{noise}). \tag{1.30}$$

We shall represent this equation in a more compact way as

$$\mathbf{f} = \mathbf{s} + \mathbf{n} \tag{1.31}$$

where $\mathbf{f}$ is the contaminated signal, $\mathbf{s}$ is the original signal, and $\mathbf{n}$ is the noise signal.

There are several kinds of noise. A few of the commonly encountered types are the following:

1. *Random noise.* The noise signal is highly oscillatory, its values alternating rapidly between values above and below an average, or mean, value. For simplicity, we shall examine random noise with a mean value of 0.

2. *Pop noise.* This type of noise is heard on old analog recordings obtained from phonograph records. The noise is perceived as randomly

occurring, isolated "pops." As a model for this type of noise we add a few non-zero values to the original signal at isolated locations.

3. *Localized random noise.* Sometimes the noise appears as in type 1, but only over a short segment or segments of the signal. This can occur when there is a short-lived disturbance in the environment during transmission of the signal.

Of course, there can also be noise signals which combine aspects of each of these types. In this section we shall examine only the first type of noise, random noise. The other types will be considered later.

Our approach will be similar to how we treated compression in the last section; we shall examine how noise removal is performed on two test signals using the Haar transform. For the first test signal, the Haar transform is used very effectively for removing the noise. For the second signal, however, the Haar transform performs poorly, and we shall need to use more sophisticated wavelet transforms to remove the noise from this signal. The essential principles, however, underlying these more sophisticated wavelet methods are the same principles we describe here for the Haar transform.

We begin by stating a basic method for removing random noise. Then we examine how this method performs on the two test signals.

Threshold Method of Wavelet Denoising

Suppose that the contaminated signal **f** equals the transmitted signal **s** plus the noise signal **n**. Also suppose that the following two conditions hold:

1. The energy of the original signal **s** is effectively captured, to a high percentage, by transform values whose magnitudes are all greater than a *threshold* $T_{\mathbf{s}} > 0$.

2. The noise signal's transform values all have magnitudes which lie below a *noise threshold* $T_{\mathbf{n}}$ satisfying $T_{\mathbf{n}} < T_{\mathbf{s}}$.

Then the noise in **f** can be removed by thresholding its transform: *All values of its transform whose magnitudes lie below the noise threshold* $T_{\mathbf{n}}$ *are set equal to* 0 *and an inverse transform is performed, providing a good approximation of* **f**.

Let's see how this method applies to Signal A shown in Figure 1.6(a). This signal was obtained by adding random noise, whose values oscillate between ± 0.1 with a mean of zero, to Signal 1 shown in Figure 1.6(a). In this case, Signal 1 is the original signal and Signal A is the contaminated signal. As we saw in the last section, the energy of Signal 1 is captured very effectively by the relatively few transform values whose magnitudes lie above a threshold of .35. So we set T_s equal to .35, and condition 1 in the Denoising Method is satisfied.

Now as for condition 2, look at the 10-level Haar transform of Signal A shown in Figure 1.6(b). Comparing this Haar transform with the Haar transform of Signal 1 in Figure 1.4(b), it is clear that the added noise has contributed a large number of small magnitude values to the transform of Signal A, while the high-energy transform values of Signal 1 are plainly visible (although slightly altered by the addition of noise). Therefore, we can satisfy condition 2 and eliminate the noise if we choose a noise threshold of, say, $T_{\mathbf{n}} = .25$. This is indicated by the two horizontal lines shown in Figure 1.6(b); all transform values lying between $\pm .25$ are set equal to 0, producing the thresholded transform shown in Figure 1.6(c). Comparing Figure 1.6(c) with Figure 1.4(b) we see that the thresholded Haar transform of the contaminated signal is a close match to the Haar transform of the original signal. Consequently, after performing an inverse transform on this thresholded signal, we obtain a denoised signal that is a close match to the original signal. This denoised signal is shown in Figure 1.6(d), and it is clearly a good approximation to Signal 1, especially considering how much noise was originally present in Signal A.

The effectiveness of noise removal can be quantitatively measured in the following way. The *Root Mean Square Error* (RMS Error) of the contaminated signal $\mathbf{f}$ compared with the original signal $\mathbf{s}$ is defined to be

$$\textit{RMS Error} = \sqrt{\frac{(f_1 - s_1)^2 + (f_2 - s_2)^2 + \cdots + (f_N - s_N)^2}{N}}. \tag{1.32}$$

Since $\mathbf{f} = \mathbf{s} + \mathbf{n}$, then $\mathbf{n} = \mathbf{f} - \mathbf{s}$. Consequently, the values of $\mathbf{n}$ are formed from the differences of the values of $\mathbf{f}$ and $\mathbf{s}$; so we can rewrite (1.32) as

$$\textit{RMS Error} = \sqrt{\frac{n_1^2 + n_2^2 + \cdots + n_N^2}{N}} = \frac{\sqrt{\mathcal{E}_{\mathbf{n}}}}{\sqrt{N}}. \tag{1.33}$$

Equation (1.33) says that the RMS Error equals the square root of the noise energy divided by $\sqrt{N}$, where N is the number of values of the signals. For example, for Signal A the RMS Error between it and Signal 1 is .057. After denoising, the RMS Error between the denoised signal and Signal 1 is .011, which shows that there is a five-fold reduction in the amount of noise. This gives quantitative evidence for the effectiveness of the denoising of Signal A.

Summarizing this example, we can say that the denoising was effective for two reasons: (1) *the transform was able to compress the energy of the original signal into a few high-energy values,* and (2) *the added noise was transformed into low-energy values.* Consequently, the high-energy transform values from the original signal stood out clearly from the low-energy noise transform values which could then be eliminated by thresholding.

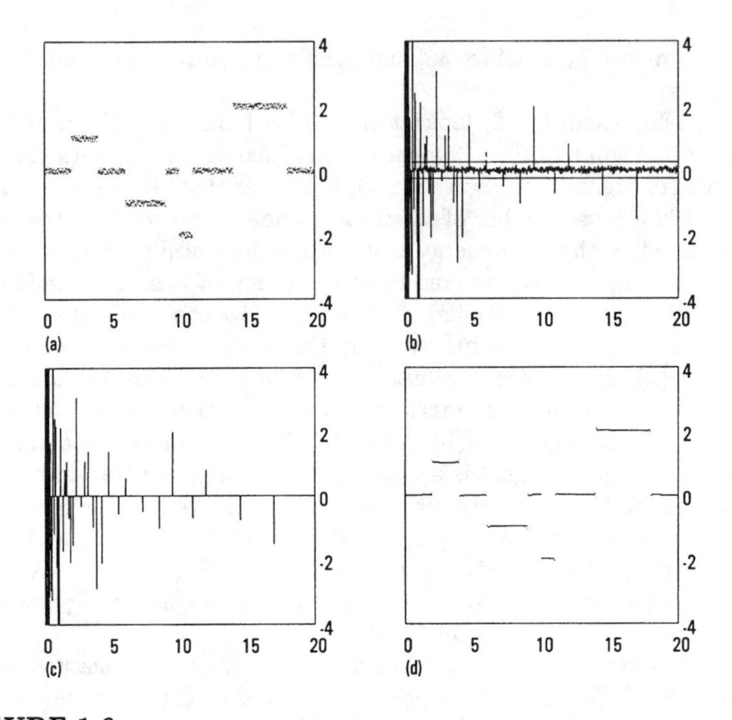

FIGURE 1.6
(a) Signal A, 2^{10} values. (b) 10-level Haar transform of Signal A. The two horizontal lines are at values of $\pm.25$ where .25 is a denoising threshold. (c) Thresholded transform. (d) Denoised signal.

Unfortunately, denoising with the Haar transform is not always so effective. Consider, for example, Signal B shown in Figure 1.7(a). This signal consists of Signal 2, shown in Figure 1.5(a), with random noise added. We view Signal 2 as the original signal and Signal B as the contaminated signal. As with the first case considered above, the random noise has values that oscillate between ± 0.1 with a mean of zero. In this case, however, we saw in the last section that it takes a relatively large number of transform values to capture the energy in Signal 2. Most of these transform values are of low energy, and it takes many of them to produce a good approximation of Signal 2. When the random noise is added to Signal 2, then the Haar transform, just like in the previous case, produces many small transform values which lie below a noise threshold. This is illustrated in Figure 1.7(b) where we show the 12-level Haar transform of Signal B. As can be seen by comparing Figure 1.7(b) with Figure 1.5(b), the small transform values that come from the noise *obscure most of the small magnitude values that result from the original signal.* Consequently, when a thresholding is done to remove the noise, as indicated by the horizontal lines in Figure 1.7(b), *this removes*

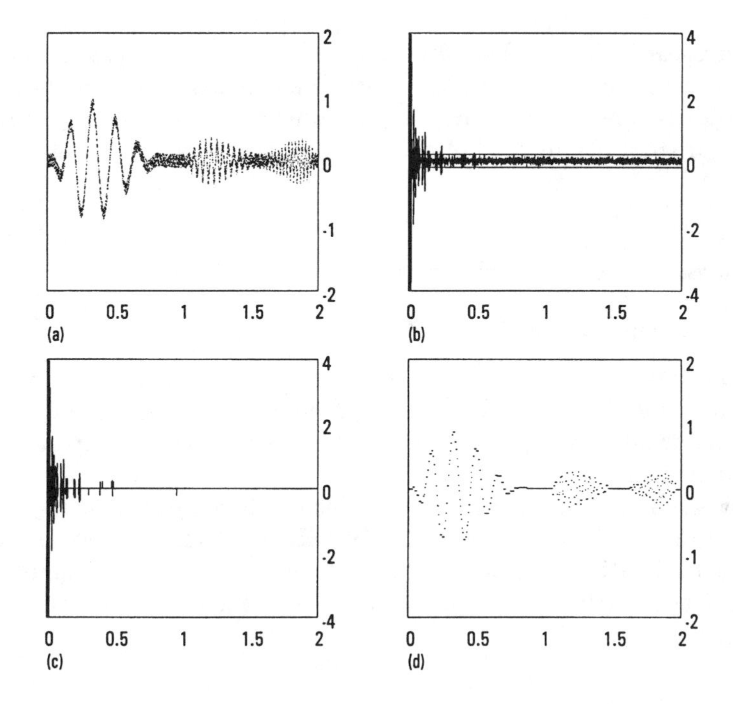

FIGURE 1.7
(a) Signal B, 2^{12} values. (b) 12-level Haar transform of Signal B. The two horizontal lines are at values of $\pm.2$ where .2 is the denoising threshold. (c) Thresholded transform. (d) Denoised signal.

many of the transform values of the original signal which are needed for an accurate approximation. This can be verified by comparing the thresholded signal shown in Figure 1.7(c) with the original signal's transform in Figure 1.5(b). In Figure 1.7(d) we show the denoised signal obtained by inverse transforming the thresholded signal. This denoised signal is clearly an unsatisfactory approximation of the original signal. By computing RMS Errors, we can quantify this judgment. The RMS Error between Signal B and Signal 2 is .057, while the RMS Error between the denoised signal and Signal 2 is .035. This shows that the error after denoising is almost two-thirds as great as the original error.

Summarizing this second test case, we can say that the denoising was not effective because *the transform could not compress the energy of the original signal into a few high-energy values lying above the noise threshold.* We shall see in the next chapter that more sophisticated wavelet transforms can achieve the desired compression and will perform nearly as well at denoising Signal B as the Haar transform did for Signal A.

We have tried to emphasize the close connection between the degree of effectiveness of the threshold denoising method and the degree of effectiveness of the wavelet transform compression method. In the next chapter we shall describe how, with more powerful wavelet transforms, a very robust and nearly optimal method of noise removal can be realized.

1.7 Notes and references

More material on the Haar transform and its applications can be found in [RAO]. Besides the lossy compression method described in this chapter, the Haar transform has also played a role in lossless image compression; see [RAJ] or [HER].

For those readers interested in the history of wavelet analysis, a good place to start is the article by Burke [BUR], which has been expanded into a book [HUB]. There is also some history, of a more sophisticated mathematical nature, in the books by Meyer [ME2] and Daubechies [DAU].

Chapter 2

Daubechies wavelets

It is hardly an exaggeration to say that we will introduce almost as many analysis algorithms as there are signals... signals are so rich and complex that a single analysis method... cannot serve them all.

Yves Meyer[1]

In this chapter we describe a large collection of wavelet transforms discovered by Daubechies. The Daubechies wavelet transforms are defined in the same way as the Haar wavelet transform—by computing running averages and differences via scalar products with scaling signals and wavelets—the only difference between them consists in how these scaling signals and wavelets are defined. For the Daubechies wavelet transforms, the scaling signals and wavelets have slightly longer supports, i.e., they produce averages and differences using just a few more values from the signal. This slight change, however, provides a tremendous improvement in the capabilities of these new transforms. They provide us with a set of powerful tools for performing basic signal processing tasks. These tasks include compression and noise removal for audio signals and for images, and include image enhancement and signal recognition.

2.1 The Daub4 wavelets

There are many Daubechies transforms, but they are all very similar. In this section we shall concentrate on the simplest one, the Daub4 wavelet

[1]Meyer's quote is from [ME2].

transform. The Daub4 wavelet transform is defined in essentially the same way as the Haar wavelet transform. If a signal $\mathbf{f}$ has an even number N of values, then the 1-level Daub4 transform is the mapping $\mathbf{f} \xrightarrow{\mathbf{D}_1} (\mathbf{a}^1 \mid \mathbf{d}^1)$ from the signal $\mathbf{f}$ to its first trend subsignal $\mathbf{a}^1$ and first fluctuation subsignal $\mathbf{d}^1$. Each value a_m of $\mathbf{a}^1 = (a_1, \ldots, a_{N/2})$ is equal to a scalar product:

$$a_m = \mathbf{f} \cdot \mathbf{V}_m^1 \tag{2.1}$$

of $\mathbf{f}$ with a 1-level *scaling signal* $\mathbf{V}_m^1$. Likewise, each value d_m of $\mathbf{d}^1 = (d_1, \ldots, d_{N/2})$ is equal to a scalar product:

$$d_m = \mathbf{f} \cdot \mathbf{W}_m^1 \tag{2.2}$$

of $\mathbf{f}$ with a 1-level *wavelet* $\mathbf{W}_m^1$. We shall define these Daub4 scaling signals and wavelets in a moment, but first we shall briefly describe the higher level Daub4 transforms.

The Daub4 wavelet transform, like the Haar transform, can be extended to multiple levels as many times as the signal length can be divided by 2. The extension is similar to the way the Haar transform is extended, i.e., by applying the 1-level Daub4 transform $\mathbf{D}_1$ to the first trend $\mathbf{a}^1$. This produces the mapping $\mathbf{a}^1 \xrightarrow{\mathbf{D}_1} (\mathbf{a}^2 \mid \mathbf{d}^2)$ from the first trend subsignal $\mathbf{a}^1$ to a second trend subsignal $\mathbf{a}^2$ and second fluctuation subsignal $\mathbf{d}^2$. The 2-level Daub4 transform $\mathbf{D}_2$ is then defined by the mapping $\mathbf{f} \xrightarrow{\mathbf{D}_2} (\mathbf{a}^2 \mid \mathbf{d}^2 \mid \mathbf{d}^1)$. For example, we show in Figure 2.2(b) the 2-level Daub4 transform of the signal shown in Figure 1.2(a). As with the Haar transform, the values of the second trend $\mathbf{a}^2$ and second fluctuation $\mathbf{d}^2$ can be obtained via scalar products with second-level scaling signals and wavelets. Likewise, the definition of a k-level Daub4 transform is obtained by applying the 1-level transform to the preceding level trend subsignal, just like in the Haar case. And, as in the Haar case, the values of the k-level trend subsignal $\mathbf{a}^k$ and fluctuation subsignal $\mathbf{d}^k$ are obtained as scalar products of the signal with k-level scaling signals and wavelets.

The difference between the Haar transform and the Daub4 transform lies in the way that the scaling signals and wavelets are defined. We shall first discuss the scaling signals. Let the *scaling numbers* $\alpha_1, \alpha_2, \alpha_3, \alpha_4$ be defined by

$$\alpha_1 = \frac{1+\sqrt{3}}{4\sqrt{2}}, \quad \alpha_2 = \frac{3+\sqrt{3}}{4\sqrt{2}}, \quad \alpha_3 = \frac{3-\sqrt{3}}{4\sqrt{2}}, \quad \alpha_4 = \frac{1-\sqrt{3}}{4\sqrt{2}}. \tag{2.3}$$

Later in this chapter and the next, we shall describe how these scaling numbers were obtained. Using these scaling numbers, the 1-level Daub4 scaling signals are

$$\begin{aligned}
\mathbf{V}_1^1 &= (\alpha_1, \alpha_2, \alpha_3, \alpha_4, 0, 0, \dots, 0) \\
\mathbf{V}_2^1 &= (0, 0, \alpha_1, \alpha_2, \alpha_3, \alpha_4, 0, 0, \dots, 0) \\
\mathbf{V}_3^1 &= (0, 0, 0, 0, \alpha_1, \alpha_2, \alpha_3, \alpha_4, 0, 0, \dots, 0) \\
&\vdots \\
\mathbf{V}_{N/2-1}^1 &= (0, 0, \dots, 0, \alpha_1, \alpha_2, \alpha_3, \alpha_4) \\
\mathbf{V}_{N/2}^1 &= (\alpha_3, \alpha_4, 0, 0, \dots, 0, \alpha_1, \alpha_2).
\end{aligned} \tag{2.4}$$

These scaling signals are all very similar to each other. For example, each scaling signal has a support of just four time-units. Notice also that the second scaling signal $\mathbf{V}_2^1$ is just a translation by two time-units of the first scaling signal $\mathbf{V}_1^1$. Likewise, the third scaling signal $\mathbf{V}_3^1$ is a translation by four time-units of $\mathbf{V}_1^1$, and each subsequent scaling signal is a translation by a multiple of two time-units of $\mathbf{V}_1^1$. There is one wrinkle here. For $\mathbf{V}_{N/2}^1$, we would have to translate $\mathbf{V}_1^1$ by $N-2$ time-units, but since $(\alpha_1, \alpha_2, \alpha_3, \alpha_4)$ has length 4 this would send α_3 and α_4 beyond the length N of the signal $\mathbf{f}$. To avoid this problem, we *wrap-around* to the start; hence $\mathbf{V}_{N/2}^1 = (\alpha_3, \alpha_4, 0, 0, \dots, 0, \alpha_1, \alpha_2)$. The Haar scaling signals also have this property of being translations by multiples of two time-units of the first scaling signal. But, since the first Haar scaling signal has a support of just two adjacent non-zero values, there is no wrap-around effect in the Haar case.

The second level Daub4 scaling signals are produced by repeating the operations that were used on the natural basis of signals $\mathbf{V}_1^0, \mathbf{V}_2^0, \dots, \mathbf{V}_N^0$ to generate the first level scaling signals.[2] Using this natural basis, the first level Daub4 scaling signals satisfy

$$\mathbf{V}_m^1 = \alpha_1 \mathbf{V}_{2m-1}^0 + \alpha_2 \mathbf{V}_{2m}^0 + \alpha_3 \mathbf{V}_{2m+1}^0 + \alpha_4 \mathbf{V}_{2m+2}^0 \tag{2.5a}$$

with a wrap-around defined by $\mathbf{V}_{n+N}^0 = \mathbf{V}_n^0$. Similarly, the second level Daub4 scaling signals are defined by

$$\mathbf{V}_m^2 = \alpha_1 \mathbf{V}_{2m-1}^1 + \alpha_2 \mathbf{V}_{2m}^1 + \alpha_3 \mathbf{V}_{2m+1}^1 + \alpha_4 \mathbf{V}_{2m+2}^1 \tag{2.5b}$$

with a wrap-around defined by $\mathbf{V}_{n+N/2}^1 = \mathbf{V}_n^1$. Notice that this wrap-around, or periodicity, of the first level scaling signals is implied by the wrap-around invoked above for the natural signal basis.

By examining Formula (2.5b), we can see that each second-level Daub4 scaling signal, $\mathbf{V}_m^2$, lives for just 10 time-units, and is a translate by $4m$ time-units of $\mathbf{V}_1^2$ (if we include wrap-around). The second-level trend values are $\{\mathbf{f} \cdot \mathbf{V}_m^2\}$ and they measure trends over 10 successive values of $\mathbf{f}$, located

[2]This natural basis of signals was defined in (1.20).

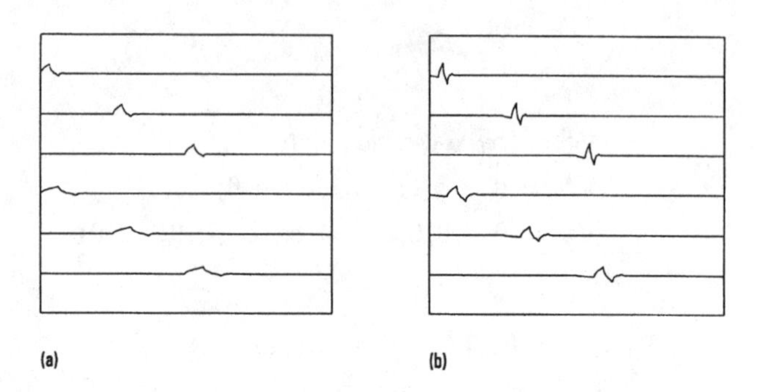

FIGURE 2.1
(a) The top 3 signals are 5-level Daub4 scaling signals $\mathbf{V}_1^5$, $\mathbf{V}_8^5$, and $\mathbf{V}_{16}^5$. The bottom three signals are 6-level scaling signals $\mathbf{V}_1^6$, $\mathbf{V}_4^6$, and $\mathbf{V}_8^6$. (b) The top 3 signals are 5-level Daub4 wavelets $\mathbf{W}_1^5$, $\mathbf{W}_8^5$, and $\mathbf{W}_{16}^5$. The bottom three signals are 6-level wavelets $\mathbf{W}_1^6$, $\mathbf{W}_4^6$, and $\mathbf{W}_8^6$.

in the same time positions as the non-zero values of $\mathbf{V}_m^2$. Hence these trends are measured over short time intervals that are shifts by multiples of 4 time-units of the interval consisting of the first 10 time-units. These 10-unit trends are slightly more than twice as long lasting as the trends measured by the first level scaling signals.

The k-level Daub4 scaling signals are defined by formulas similar to (2.5a) and (2.5b), but applied to the preceding level scaling signals. In Figure 2.1(a), we show some 5-level and 6-level Daub4 scaling signals. Notice that the supports of the 6-level scaling signals are about twice as long as the supports of the 5-level scaling signals. Figure 2.1(a) also illustrates the fact that each of the 5-level scaling signals is a translate of $\mathbf{V}_1^5$, and that each of the 6-level scaling signals is a translate of $\mathbf{V}_1^6$.

An important property of these scaling signals is that they all have energy 1. This is because of the following identity satisfied by the scaling numbers:

$$\alpha_1^2 + \alpha_2^2 + \alpha_3^2 + \alpha_4^2 = 1. \tag{2.6}$$

It is clear that (2.6) implies that each 1-level scaling signal has energy 1. To see that it also implies that each k-level scaling signal has energy 1 is more difficult; we will sketch the proof at the end of the next section.

Another identity satisfied by the scaling numbers is

$$\alpha_1 + \alpha_2 + \alpha_3 + \alpha_4 = \sqrt{2}. \tag{2.7}$$

This equation says that each 1-level trend value $\mathbf{f} \cdot \mathbf{V}_m^1$ is an average of four values of $\mathbf{f}$, multiplied by $\sqrt{2}$. It can also be shown that the sum of the ten successive non-zero values of $\mathbf{V}_m^2$ is 2, which shows that each 2-level trend value $\mathbf{f} \cdot \mathbf{V}_m^2$ is an average of ten successive values of $\mathbf{f}$, multiplied by 2.

Similarly, each k-level trend value $\mathbf{f} \cdot \mathbf{V}_m^k$ is an average of values of $\mathbf{f}$ over increasingly longer time intervals as k increases.

We now turn to a discussion of the Daub4 wavelets. Let the *wavelet numbers* $\beta_1, \beta_2, \beta_3, \beta_4$ be defined by

$$\beta_1 = \frac{1-\sqrt{3}}{4\sqrt{2}}, \quad \beta_2 = \frac{\sqrt{3}-3}{4\sqrt{2}}, \quad \beta_3 = \frac{3+\sqrt{3}}{4\sqrt{2}}, \quad \beta_4 = \frac{-1-\sqrt{3}}{4\sqrt{2}}. \tag{2.8}$$

Notice that the wavelet numbers are related to the scaling numbers by the equations: $\beta_1 = \alpha_4$, $\beta_2 = -\alpha_3$, $\beta_3 = \alpha_2$, and $\beta_4 = -\alpha_1$. Using these wavelet numbers, the 1-level Daub4 wavelets $\mathbf{W}_1^1, \ldots, \mathbf{W}_{N/2}^1$ are defined by

$$\begin{aligned}
\mathbf{W}_1^1 &= (\beta_1, \beta_2, \beta_3, \beta_4, 0, 0, \ldots, 0) \\
\mathbf{W}_2^1 &= (0, 0, \beta_1, \beta_2, \beta_3, \beta_4, 0, 0, \ldots, 0) \\
\mathbf{W}_3^1 &= (0, 0, 0, 0, \beta_1, \beta_2, \beta_3, \beta_4, 0, 0, \ldots, 0) \\
&\vdots \\
\mathbf{W}_{N/2-1}^1 &= (0, 0, \ldots, 0, \beta_1, \beta_2, \beta_3, \beta_4) \\
\mathbf{W}_{N/2}^1 &= (\beta_3, \beta_4, 0, 0, \ldots, 0, \beta_1, \beta_2).
\end{aligned} \tag{2.9}$$

These wavelets are all translates of $\mathbf{W}_1^1$, with a wrap-around for the last wavelet. Each wavelet has a support of just four time-units, corresponding to the four non-zero wavelet numbers used to define them. The 1-level Daub4 wavelets satisfy

$$\mathbf{W}_m^1 = \beta_1 \mathbf{V}_{2m-1}^0 + \beta_2 \mathbf{V}_{2m}^0 + \beta_3 \mathbf{V}_{2m+1}^0 + \beta_4 \mathbf{V}_{2m+2}^0.$$

Similarly, the 2-level Daub4 wavelets are defined by

$$\mathbf{W}_m^2 = \beta_1 \mathbf{V}_{2m-1}^1 + \beta_2 \mathbf{V}_{2m}^1 + \beta_3 \mathbf{V}_{2m+1}^1 + \beta_4 \mathbf{V}_{2m+2}^1.$$

All other levels of Daub4 wavelets are defined in a similar fashion. In Figure 2.1(b) we show some of the Daub4 wavelets. It is interesting to compare them with the Daub4 scaling functions shown in Figure 2.1(a).

The Daub4 wavelets all have energy 1. This is clear for the 1-level Daub4 wavelets, since

$$\beta_1^2 + \beta_2^2 + \beta_3^2 + \beta_4^2 = 1. \tag{2.10}$$

It can also be shown that all k-level Daub4 wavelets have energy 1 as well.

Each fluctuation value $d_m = \mathbf{f} \cdot \mathbf{W}_m^1$ can be viewed as a differencing operation on the values of $\mathbf{f}$ because

$$\beta_1 + \beta_2 + \beta_3 + \beta_4 = 0. \tag{2.11}$$

Equation (2.11) is a generalization of the Haar case, where we had $1/\sqrt{2} - 1/\sqrt{2} = 0$. It also implies, as with the Haar case, that a fluctuation value

$\mathbf{f} \cdot \mathbf{W}_m^1$ will be zero if the signal $\mathbf{f}$ is constant over the support of a Daub4 wavelet $\mathbf{W}_m^1$. Much more is true, however. Not only is (2.11) true, but we also have

$$0\beta_1 + 1\beta_2 + 2\beta_3 + 3\beta_4 = 0. \tag{2.12}$$

Equations (2.11) and (2.12), and Equation (2.7), imply the following property for the k-level Daub4 wavelet transform.

Property I. *If a signal* $\mathbf{f}$ *is (approximately) linear over the support of a* k*-level Daub4 wavelet* $\mathbf{W}_m^k$*, then the* k*-level fluctuation value* $\mathbf{f} \cdot \mathbf{W}_m^k$ *is (approximately) zero.*

For the 1-level case, Property I follows easily from Equations (2.11) and (2.12). It is more difficult to prove Property I for the k-level case, when $k > 1$, and it is for such cases that Equation (2.7) is needed.

To see why Property I is so important, we examine how it relates to sampled signals. In Figure 2.2(a) we show a signal obtained from uniformly spaced samples over the interval $[0, 1)$ of a function which has a continuous second derivative. As shown in Figure 2.2(b), the 1-level and 2-level fluctuations $\mathbf{d}^1$ and $\mathbf{d}^2$ have values that are all very near zero. This is because a large proportion of the signal consists of values that are approximately linear over a support of one of the Daub4 wavelets. For example, in Figures 2.2(c) and (d) we show magnifications of small squares centered at the points $(.296, .062)$ and $(.534, .067)$. It is clear from these figures that the signal values are approximately linear within these small squares. This is true of a large number of points on the graph of the signal and implies that many of the fluctuation values for this signal will be near zero. The basic principles of Calculus tell us that this example is typical for a signal that is sampled from a function that has a continuous second derivative. We shall discuss this point in more detail at the end of this section.

Each level Daub4 transform has an inverse. The inverse for the 1-level Daub4 transform, which maps the transform $(\mathbf{a}^1 \,|\, \mathbf{d}^1)$ back to the signal $\mathbf{f}$, is calculated explicitly as

$$\mathbf{f} = \mathbf{A}^1 + \mathbf{D}^1 \tag{2.13a}$$

with *first averaged signal* $\mathbf{A}^1$ defined by

$$\begin{aligned} \mathbf{A}^1 &= a_1\mathbf{V}_1^1 + a_2\mathbf{V}_2^1 + \cdots + a_{N/2}\mathbf{V}_{N/2}^1 \\ &= (\mathbf{f}\cdot\mathbf{V}_1^1)\mathbf{V}_1^1 + (\mathbf{f}\cdot\mathbf{V}_2^1)\mathbf{V}_2^1 + \cdots + (\mathbf{f}\cdot\mathbf{V}_{N/2}^1)\mathbf{V}_{N/2}^1 \end{aligned} \tag{2.13b}$$

and *first detail signal* $\mathbf{D}^1$ defined by

$$\begin{aligned} \mathbf{D}^1 &= d_1\mathbf{W}_1^1 + d_2\mathbf{W}_2^1 + \cdots + d_{N/2}\mathbf{W}_{N/2}^1 \\ &= (\mathbf{f}\cdot\mathbf{W}_1^1)\mathbf{W}_1^1 + (\mathbf{f}\cdot\mathbf{W}_2^1)\mathbf{W}_2^1 + \cdots + (\mathbf{f}\cdot\mathbf{W}_{N/2}^1)\mathbf{W}_{N/2}^1. \end{aligned} \tag{2.13c}$$

Formulas (2.13a) through (2.13c) are generalizations of similar formulas that we found for the Haar transform [see Section 1.4]. They are the first stage in a Daub4 MRA of the signal $\mathbf{f}$. We will not take the time at this point to prove that these formulas are correct; their proof involves techniques from the field of linear algebra. For those readers who are familiar with linear algebra, we provide a proof at the end of the next section. It is more important now to reflect on what these formulas mean.

Formula (2.13a) shows that the signal $\mathbf{f}$ can be expressed as a sum of an averaged signal $\mathbf{A}^1$ plus a detail signal $\mathbf{D}^1$. Because of Formula (2.13b) we can see that the averaged signal $\mathbf{A}^1$ is a combination of elementary scaling signals. Each scaling signal $\mathbf{V}_m^1$ is a short-lived signal, whose support consists of just four consecutive time-indices; the relative contribution of each scaling signal to $\mathbf{A}^1$ is measured by the trend value $a_m = \mathbf{f} \cdot \mathbf{V}_m^1$. Thus $\mathbf{A}^1$ is a sum of short-lived components which are multiples of the scaling signals $\mathbf{V}_m^1$. These scaling signals move across the time-axis in steps of just two time-units and live for only four time-units; *they measure short-lived trends in the signal via the trend values* $a_m = \mathbf{f} \cdot \mathbf{V}_m^1$. Likewise, the detail signal $\mathbf{D}^1$ is a combination of elementary wavelets $\mathbf{W}_m^1$. These wavelets $\mathbf{W}_m^1$ march across the time-axis in steps of two time-units and live for only four time-units. The relative contribution of each wavelet to $\mathbf{D}^1$ is measured by the fluctuation value $d_m = \mathbf{f} \cdot \mathbf{W}_m^1$. Since $\mathbf{D}^1 = \mathbf{f} - \mathbf{A}^1$, the sum of all of these short-lived fluctuations $\{d_m \mathbf{W}_m^1\}$ equals the difference between the signal $\mathbf{f}$ and its lower resolution, averaged, version $\mathbf{A}^1$. Because the 1-level wavelets $\{\mathbf{W}_m^1\}$ live for only four time-units and march across the time-axis in steps of two units, *they are able to detect very short-lived, transient, fluctuations in the signal.*

The inverse of the 2-level Daub4 transform is described by the formula

$$\mathbf{f} = \mathbf{A}^2 + \mathbf{D}^2 + \mathbf{D}^1 \tag{2.14a}$$

where

$$\mathbf{A}^2 = (\mathbf{f} \cdot \mathbf{V}_1^2)\mathbf{V}_1^2 + \cdots + (\mathbf{f} \cdot \mathbf{V}_{N/4}^2)\mathbf{V}_{N/4}^2$$

$$\mathbf{D}^2 = (\mathbf{f} \cdot \mathbf{W}_1^2)\mathbf{W}_1^2 + \cdots + (\mathbf{f} \cdot \mathbf{W}_{N/4}^2)\mathbf{W}_{N/4}^2 \tag{2.14b}$$

are the second averaged signal and second detail signal, respectively. The signal $\mathbf{D}^1$ is the first detail signal which we defined above. The second averaged signal $\mathbf{A}^2$ is a sum of components which are multiples of the scaling signals $\mathbf{V}_m^2$; these components move across the time-axis in steps of four time-units and live for ten time-units. The relative contribution of each scaling signal $\mathbf{V}_m^2$ to $\mathbf{A}^2$ is measured by the trend value $\mathbf{f} \cdot \mathbf{V}_m^2$. Since $\mathbf{A}^1 = \mathbf{A}^2 + \mathbf{D}^2$, the second detail signal $\mathbf{D}^2$ provides the details needed to produce the first averaged signal from the second averaged signal. This second detail signal $\mathbf{D}^2$ is a combination of the wavelets $\mathbf{W}_m^2$, which move

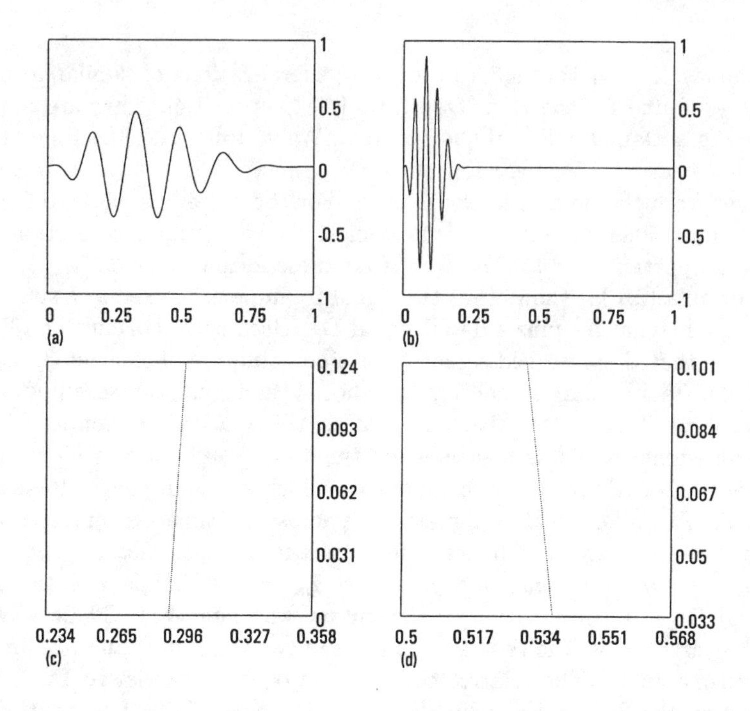

FIGURE 2.2
(a) Signal. (b) 2-level Daub4 transform. The trend $\mathbf{a}^2$ is graphed over [0, .25), while the fluctuations $\mathbf{d}^2$ and $\mathbf{d}^1$ are graphed over [.25, .5) and [.5, 1), respectively. (c) and (d) Magnifications of the Signal's graph in two small squares; the Signal is approximately linear.

across the time-axis in steps of four time-units and live for ten time-units. The relative contribution of each wavelet $\mathbf{W}_m^2$ to $\mathbf{D}^2$ is measured by the fluctuation value $\mathbf{f} \cdot \mathbf{W}_m^2$. Like the 1-level wavelets, the 2-level wavelets are also able to detect transient fluctuations in a signal, but their supports are 10 units long instead of 4 units long. Hence the scale on which the 2-level wavelets measure fluctuations is slightly more than twice as long as the scale on which the 1-level wavelets measure fluctuations.

Further levels of the Daub4 transform are handled in a like manner. The k-level Daub4 transform has an inverse that produces the following MRA of the signal $\mathbf{f}$:

$$\mathbf{f} = \mathbf{A}^k + \mathbf{D}^k + \cdots + \mathbf{D}^2 + \mathbf{D}^1.$$

The formulas for $\mathbf{A}^k$ and $\mathbf{D}^k$ are (for $N_k = N/2^k$):

$$\mathbf{A}^k = (\mathbf{f} \cdot \mathbf{V}_1^k)\mathbf{V}_1^k + \cdots + (\mathbf{f} \cdot \mathbf{V}_{N_k}^k)\mathbf{V}_{N_k}^k$$

and

$$\mathbf{D}^k = (\mathbf{f} \cdot \mathbf{W}_1^k)\mathbf{W}_1^k + \cdots + (\mathbf{f} \cdot \mathbf{W}_{N_k}^k)\mathbf{W}_{N_k}^k.$$

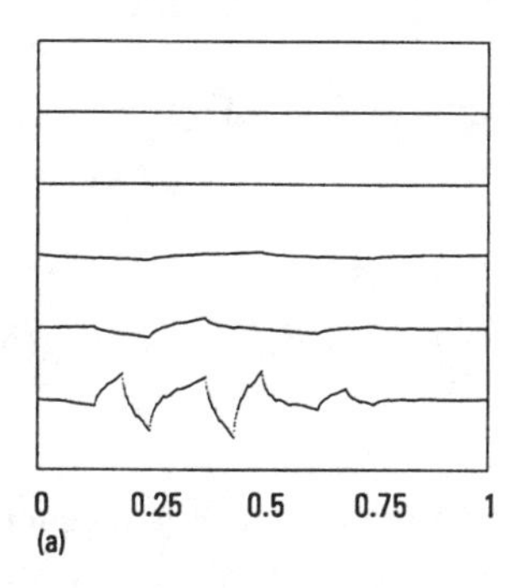

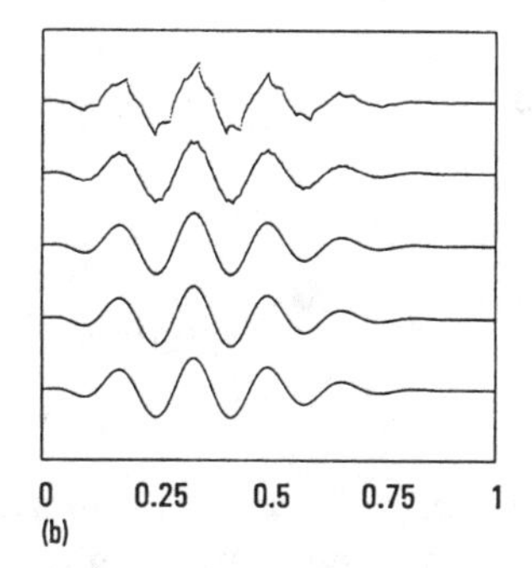

FIGURE 2.3
Daub4 MRA of the signal shown in Figure 1.1(a). The graphs are of the 10 averaged signals $\mathbf{A}^{10}$ through $\mathbf{A}^1$. Beginning with $\mathbf{A}^{10}$ on the top left down to $\mathbf{A}^6$ on the bottom left, then $\mathbf{A}^5$ on the top right down to $\mathbf{A}^1$ on the bottom right. Compare with Figure 1.3.

In Figure 2.3 we show a Daub4 MRA of the same signal that we analyzed in Chapter 1 using a Haar MRA (see Figure 1.3). It is interesting to compare these two MRAs. The Daub4 MRA appears to be the superior one; it converges more quickly towards the original signal. The Daub4 averaged signals $\mathbf{A}^3$ through $\mathbf{A}^1$ all appear to be equally close approximations of the original signal. This is due to the fact that the values of the first and second Daub4 fluctuation subsignals $\mathbf{d}^1$ and $\mathbf{d}^2$ are so small [see Figure 2.2(b)] that they can be neglected without losing much detail. Likewise, the third Daub4 fluctuation subsignal $\mathbf{d}^3$ has negligible values. The corresponding Daub4 detail signals $\mathbf{D}^1$, $\mathbf{D}^2$, and $\mathbf{D}^3$ contribute very little detail to the signal; hence $\mathbf{f} \approx \mathbf{A}^3$ is a very good approximation. Another advantage of the Daub4 MRA is that the jumpy, or clumpy, appearance of the Haar averaged signals does not appear in the Daub4 averaged signals.

Further remarks on Property I*

In discussing Property I above, we showed by means of an example that it applies to sampled signals when the analog signal has a continuous second derivative over the support of a Daub4 wavelet. That is, we assume that the signal $\mathbf{f}$ has values satisfying $f_n = g(t_n)$ for $n = 1, 2, \ldots, N$, and that the function g has a continuous second derivative over the support of a Daub4 wavelet. For simplicity we shall assume that this is a 1-level wavelet, say $\mathbf{W}_m^1$. We can then write

$$g(t_{2m-1+k}) = g(t_{2m-1}) + g'(t_{2m-1})(kh) + \mathrm{O}(h^2) \qquad (2.15)$$

where $\mathrm{O}(h^2)$ stands for a quantity that is a bounded multiple of h^2. The number h is the constant step-size $h = t_{n+1} - t_n$, which holds for each n.

Making use of Equation (2.15), and Equations (2.11) and (2.12), we find that

$$\begin{aligned}\mathbf{f}\cdot\mathbf{W}_m^1 &= g(t_{2m-1})\{\beta_1+\beta_2+\beta_3+\beta_4\}\\ &\quad +g'(t_{2m-1})h\{0\beta_1+1\beta_2+2\beta_3+3\beta_4\}+\mathrm{O}(h^2)\\ &= \mathrm{O}(h^2).\end{aligned}$$

Thus $\mathbf{f}\cdot\mathbf{W}_m^1 = \mathrm{O}(h^2)$. This illustrates Property I, since h is generally much smaller than 1 and consequently h^2 is very tiny indeed. Our discussion also shows why the Daub4 transform generally produces much smaller fluctuation values than the Haar transform does, since for the Haar transform it is typically possible only to have $\mathbf{f}\cdot\mathbf{W}_m^1 = \mathrm{O}(h)$, which is generally much larger than $\mathrm{O}(h^2)$.

2.2 Conservation and compaction of energy

Like the Haar transform, a Daubechies wavelet transform conserves the energy of signals and redistributes this energy into a more compact form. In this section we shall discuss these properties as they relate to the Daub4 transform, but the general principles apply to all of the various Daubechies transforms.

Let's begin with a couple of examples. First, consider the signal $\mathbf{f}$ graphed in Figure 2.2(a). Using FAWAV we calculate that its energy is 509.2394777. On the other hand, the energy of its 1-level Daub4 transform is also 509.2394777. This illustrates the conservation of energy property of the Daub4 transform. Since the 2-level Daub4 transform consists of applying the 1-level Daub4 transform to the first trend subsignal, it follows that the 2-level Daub4 transform also conserves the energy of the signal $\mathbf{f}$. Likewise, a k-level Daub4 transform conserves energy as well.

As with the Haar transform, the Daub4 transform also redistributes the energy of the signal into a more compact form. For example, consider the signals shown in Figure 2.4. In Figure 2.4(b) we show the 2-level Daub4 transform of the signal graphed in Figure 1.2(a). Its cumulative energy profile is graphed in Figure 2.4(d). This cumulative energy profile shows that the 2-level Daub4 transform has redistributed almost all of the energy of the signal into the second trend subsignal, which is graphed over the first quarter of the time-interval. For comparison, we also show in Figures 2.4(a) and (c) the 2-level Haar transform and its cumulative energy profile. It is obvious from these graphs that the Daub4 transform achieves a more compact redistribution of the energy of the signal.

Justification of conservation of energy *

We shall now show why the Daub4 transform preserves the energy of each signal, and provide justifications for a couple of other statements made in the preceding section. Readers who are not familiar with linear algebra, especially matrix algebra, should skip this discussion. It will not play a major role in the material that follows, which will stress the applications of the Daubechies transforms.

To begin, define the matrix $\mathcal{D}_N$ by

$$\mathcal{D}_N = \begin{pmatrix} \alpha_1 & \alpha_2 & \alpha_3 & \alpha_4 & 0 & 0 & 0 & \dots & 0 & 0 & 0 \\ \beta_1 & \beta_2 & \beta_3 & \beta_4 & 0 & 0 & 0 & \dots & 0 & 0 & 0 \\ 0 & 0 & \alpha_1 & \alpha_2 & \alpha_3 & \alpha_4 & 0 & \dots & 0 & 0 & 0 \\ 0 & 0 & \beta_1 & \beta_2 & \beta_3 & \beta_4 & 0 & \dots & 0 & 0 & 0 \\ \vdots & \vdots & \vdots & \vdots & \vdots & \vdots & \vdots & \dots & \vdots & \vdots & \vdots \\ \alpha_3 & \alpha_4 & 0 & 0 & 0 & 0 & 0 & \dots & 0 & \alpha_1 & \alpha_2 \\ \beta_3 & \beta_4 & 0 & 0 & 0 & 0 & 0 & \dots & 0 & \beta_1 & \beta_2 \end{pmatrix}. \tag{2.16}$$

Notice that the rows of $\mathcal{D}_N$ are the first-level Daub4 scaling signals and wavelets. These scaling signals and wavelets satisfy

$$\mathbf{V}_n^1 \cdot \mathbf{V}_m^1 = \begin{cases} 1 & \text{if } n = m \\ 0 & \text{if } n \neq m, \end{cases} \tag{2.17a}$$

$$\mathbf{W}_n^1 \cdot \mathbf{W}_m^1 = \begin{cases} 1 & \text{if } n = m \\ 0 & \text{if } n \neq m, \end{cases} \tag{2.17b}$$

$$\mathbf{V}_n^1 \cdot \mathbf{W}_m^1 = 0 \quad \text{all } m, n. \tag{2.17c}$$

These equations show that the rows of $\mathcal{D}_N$ form an orthonormal set of vectors, i.e., that $\mathcal{D}_N$ is an orthogonal matrix. Another way of expressing these equations is

$$\mathcal{D}_N^{\mathrm{T}} \mathcal{D}_N = I_N \tag{2.18}$$

where $\mathcal{D}_N^{\mathrm{T}}$ is the transpose of the matrix $\mathcal{D}_N$ and I_N is the N by N identity matrix.

We can now show that the Daub4 transform preserves the energy of a signal $\mathbf{f}$. *These arguments will only make use of Equations (2.17a) to (2.17c), and (2.18). Therefore they will apply to all of the Daubechies transforms described in this chapter, since all of the Daubechies scaling signals and wavelets will satisfy these same equations.* The matrix $\mathcal{D}_N$ will, in each case, be defined by rows consisting of the 1-level scaling signals and wavelets.

Comparing the definition of the matrix $\mathcal{D}_N$ and the definition of the 1-level Daub4 transform, we see that

$$(a_1, d_1, a_2, d_2, \dots, a_{N/2}, d_{N/2})^{\mathrm{T}} = \mathcal{D}_N \, \mathbf{f}^{\mathrm{T}}.$$

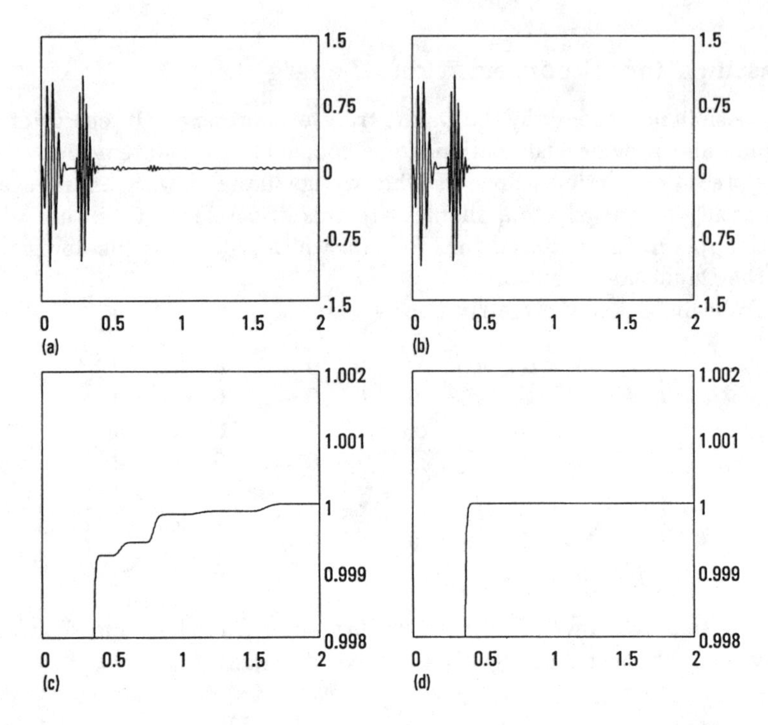

FIGURE 2.4
(a) Graph of 2-level Haar transform of signal in Figure 1.2(a). (b) Graph of 2-level Daub4 transform of same signal. (c) and (d) Cumulative energy profiles of the transforms in (a) and (b), respectively.

Therefore,

$$a_1^2 + d_1^2 + \cdots + a_{N/2}^2 + d_{N/2}^2 = \left(\mathcal{D}_N \, \mathbf{f}^{\mathrm{T}}\right)^{\mathrm{T}} \left(\mathcal{D}_N \, \mathbf{f}^{\mathrm{T}}\right).$$

Furthermore, the energy $\mathcal{E}_{(\mathbf{a}^1 \,|\, \mathbf{d}^1)}$ of the 1-level Daub4 transform of $\mathbf{f}$ satisfies

$$a_1^2 + \cdots + a_{N/2}^2 + d_1^2 + \cdots + d_{N/2}^2 = \mathcal{E}_{(\mathbf{a}^1 \,|\, \mathbf{d}^1)}.$$

Since the left-hand sides of these last two equations are clearly equal, we make use of (2.18) to obtain

$$\begin{aligned}
\mathcal{E}_{(\mathbf{a}^1 \,|\, \mathbf{d}^1)} &= (\mathcal{D}_N \mathbf{f}^{\mathrm{T}})^{\mathrm{T}} \, (\mathcal{D}_N \mathbf{f}^{\mathrm{T}}) \\
&= \mathbf{f} \, \mathcal{D}_N^{\mathrm{T}} \, \mathcal{D}_N \, \mathbf{f}^{\mathrm{T}} \\
&= \mathbf{f} \, \mathbf{f}^{\mathrm{T}} \\
&= \mathcal{E}_{\mathbf{f}}.
\end{aligned}$$

This proves that the 1-level Daub4 transform has the Conservation of Energy property. As we argued above, this also shows that every level Daub4 transform conserves energy.

Another consequence of Equations (2.17a) to (2.17c) is that the Daub4 scaling signals and wavelets all have energy 1. Since the energy $\mathcal{E}_{\mathbf{f}}$ of a signal equals $\mathbf{f} \cdot \mathbf{f}$, these equations show immediately that $\mathbf{V}_m^1$ and $\mathbf{W}_m^1$ each have energy 1. To indicate why all other scaling signals and wavelets have energy 1, we will show why the wavelet $\mathbf{W}_1^2$ has energy 1. Similar arguments can be used for the other wavelets and scaling signals. Since

$$\mathbf{W}_1^2 = \beta_1 \mathbf{V}_1^1 + \beta_2 \mathbf{V}_2^1 + \beta_3 \mathbf{V}_3^1 + \beta_4 \mathbf{V}_4^1$$

we find, using (2.17a), that

$$\mathbf{W}_1^2 \cdot \mathbf{W}_1^2 = \beta_1^2 \mathbf{V}_1^1 \cdot \mathbf{V}_1^1 + \beta_2^2 \mathbf{V}_2^1 \cdot \mathbf{V}_2^1 + \beta_3^2 \mathbf{V}_3^1 \cdot \mathbf{V}_3^1 + \beta_4^2 \mathbf{V}_4^1 \cdot \mathbf{V}_4^1. \tag{2.19}$$

Notice that terms such as $\beta_n \beta_m \mathbf{V}_n^1 \cdot \mathbf{V}_m^1$ are equal to 0 when $m \neq n$; so they do not appear on the right-hand side of (2.19). Each scalar product on the right-hand side of (2.19) is 1; hence

$$\mathbf{W}_1^2 \cdot \mathbf{W}_1^2 = \beta_1^2 + \beta_2^2 + \beta_3^2 + \beta_4^2 = 1$$

and that proves that the energy of $\mathbf{W}_1^2$ is 1.

Similar arguments can be used to show that Equations (2.13a) through (2.13c) are valid. We shall briefly indicate why this is so. Suppose that a signal $\mathbf{f}$ is defined by

$$r_1 \mathbf{V}_1^1 + \cdots + r_{N/2} \mathbf{V}_{N/2}^1 + s_1 \mathbf{W}_1^1 + \cdots + s_{N/2} \mathbf{W}_{N/2}^1 = \mathbf{f} \tag{2.20}$$

where $r_1, \ldots, r_{N/2}, s_1, \ldots, s_{N/2}$ are constants. Then, by using Equations (2.17a) through (2.17c), it follows that

$$r_m = \mathbf{f} \cdot \mathbf{V}_m^1, \quad s_m = \mathbf{f} \cdot \mathbf{W}_m^1 \tag{2.21}$$

for each m. In particular, if $\mathbf{f} = (0, 0, \ldots, 0)$, then $r_1 = 0, r_2 = 0, \ldots, r_{N/2} = 0$ and $s_1 = 0, s_2 = 0, \ldots, s_{N/2} = 0$. This proves that the signals

$$\mathbf{V}_1^1, \ldots, \mathbf{V}_{N/2}^1, \mathbf{W}_1^1, \ldots, \mathbf{W}_{N/2}^1$$

are linearly independent; hence they form a basis for the vector space $\mathbf{R}^N$ of all real-valued signals of length N. Consequently Equation (2.20) must hold, with unique coefficients $r_1, \ldots, r_{N/2}, s_1, \ldots, s_{N/2}$, for every signal $\mathbf{f}$. And Formula (2.21) shows that these coefficients are equal to the scalar products of $\mathbf{f}$ with the scaling signals and wavelets, exactly as described in Equations (2.13a) through (2.13c).

How wavelet and scaling numbers are found *

In this subsection we shall briefly outline how the Daub4 scaling numbers and wavelet numbers are determined. The essential features of this outline also apply to the other Daubechies wavelets that are defined in the next section.

The constraints that determine the Daub4 scaling and wavelet numbers are Equations (2.17a) to (2.17c), (2.6), (2.7), and (2.10) through (2.12). By combining these last three equations with the requirement that $\mathbf{W}_1^1 \cdot \mathbf{W}_2^1 = 0$ from (2.17b), we obtain the following four constraining equations on the wavelet numbers:

$$\begin{aligned} \beta_1^2 + \beta_2^2 + \beta_3^2 + \beta_4^2 &= 1 \\ \beta_1 + \beta_2 + \beta_3 + \beta_4 &= 0 \\ 0\beta_1 + 1\beta_2 + 2\beta_3 + 3\beta_4 &= 0 \\ \beta_1\beta_3 + \beta_2\beta_4 &= 0. \end{aligned}$$

These equations are sufficient for uniquely determining (except for multiplication by -1) the values for β_1, β_2, β_3, and β_4 for the Daub4 wavelet numbers. The scaling numbers are then determined by the equations $\alpha_1 = -\beta_4$, $\alpha_2 = \beta_3$, $\alpha_3 = -\beta_2$, and $\alpha_4 = \beta_1$.

This very brief outline only partially answers the question of how scaling numbers and wavelet numbers are found. We shall provide a more complete discussion in the next chapter.

2.3 Other Daubechies wavelets

In this section we shall complete our introduction to the theory of the Daubechies wavelets and wavelet transforms. In the previous two sections we described the Daub4 wavelet transform and its associated set of scaling signals and wavelets. We shall now complete our discussion by describing the various Daub$J$ transforms for $J = 6, 8, \ldots, 20$, and by describing the Coif$I$ transforms for $I = 6, 12, 18, 24, 30$. These wavelet transforms are all quite similar to the Daub4 transform; our treatment here will concentrate on the value of having more wavelet transforms at our disposal. There are also many more wavelet transforms—such as *spline wavelet transforms,* various types of *biorthogonal wavelet transforms,* and even more Daub$J$ and Coif$I$ transforms than the ones we describe—but we shall not try to give an exhaustive coverage of all of these transforms. Examining a few trees should give us a good feeling for the forest of wavelet transforms.

Let's begin with the Daub$J$ transforms for $J = 6, 8, \ldots, 20$. The easiest way to understand these transforms is just to treat them as simple generalizations of the Daub4 transform. The most obvious difference between them is the length of the supports of their scaling signals and wavelets. For example, for the Daub6 wavelet transform, we define the scaling numbers $\alpha_1, \ldots, \alpha_6$ to be

$$\alpha_1 = 0.332670552950083, \qquad \alpha_2 = 0.806891509311092,$$
$$\alpha_3 = 0.459877502118491, \qquad \alpha_4 = -0.135011020010255,$$
$$\alpha_5 = -0.00854412738820267, \quad \alpha_6 = 0.00352262918857095$$

and the wavelet numbers $\beta_1, \ldots, \beta_6$ to be

$$\beta_1 = \alpha_6,\ \beta_2 = -\alpha_5,\ \beta_3 = \alpha_4,\ \beta_4 = -\alpha_3,\ \beta_5 = \alpha_2,\ \beta_6 = -\alpha_1.$$

We then generalize the formulas in (2.4) in the following way:

$$\begin{aligned}
\mathbf{V}_1^1 &= (\alpha_1, \alpha_2, \alpha_3\, a_4, \alpha_5, \alpha_6, 0, 0, \ldots, 0)\\
\mathbf{V}_2^1 &= (0, 0, \alpha_1, \alpha_2, \alpha_3\, a_4, \alpha_5, \alpha_6, 0, 0, \ldots, 0)\\
\mathbf{V}_3^1 &= (0, 0, 0, 0, \alpha_1, \alpha_2, \alpha_3\, a_4, \alpha_5, \alpha_6, 0, 0, \ldots, 0)\\
&\ \vdots\\
\mathbf{V}_{N/2}^1 &= (\alpha_3\, a_4, \alpha_5, \alpha_6, 0, 0, \ldots, 0, \alpha_1, a_2)
\end{aligned} \tag{2.22}$$

with a wrap-around occurring for $\mathbf{V}_{N/2-1}^1$ and $\mathbf{V}_{N/2}^1$. The formulas in (2.22) define the first level Daub6 scaling signals. The scaling numbers satisfy (to a high degree of accuracy):

$$\alpha_1^2 + \alpha_2^2 + \alpha_3^2 + \alpha_4^2 + \alpha_5^2 + \alpha_6^2 = 1, \tag{2.23a}$$
$$\alpha_1 + \alpha_2 + \alpha_3 + \alpha_4 + \alpha_5 + \alpha_6 = \sqrt{2}. \tag{2.23b}$$

Equation (2.23a) says that each scaling signal $\mathbf{V}_m^1$ has an energy of 1, while Equation (2.23b) says that the trend values $\mathbf{f} \cdot \mathbf{V}_m^1$ are averages of six successive values of $\mathbf{f}$, multiplied by $\sqrt{2}$.

The 1-level Daub6 wavelets are defined via the wavelet numbers $\beta_1, \ldots, \beta_6$ in the same manner. In fact, since all of the definitions and formulas given in the last two sections generalize in obvious ways, we shall not repeat them.

Let's consider instead what new features are exhibited by the Daub6 transform. The principal feature is that the wavelet numbers $\beta_1, \ldots, \beta_6$ satisfy the following three identities (to a high degree of accuracy):

$$\begin{aligned}
\beta_1 + \beta_2 + \beta_3 + \beta_4 + \beta_5 + \beta_6 &= 0,\\
0\beta_1 + 1\beta_2 + 2\beta_3 + 3\beta_4 + 4\beta_5 + 5\beta_6 &= 0,\\
0^2\beta_1 + 1^2\beta_2 + 2^2\beta_3 + 3^2\beta_4 + 4^2\beta_5 + 5^2\beta_6 &= 0.
\end{aligned} \tag{2.24}$$

These equations, along with Equation (2.23b), imply the following property.

Property II. *If a signal* $\mathbf{f}$ *is (approximately) quadratic over the support of a* k*-level Daub6 wavelet* $\mathbf{W}_m^k$, *then the* k*-level Daub6 fluctuation value* $\mathbf{f} \cdot \mathbf{W}_m^k$ *is (approximately) zero.*

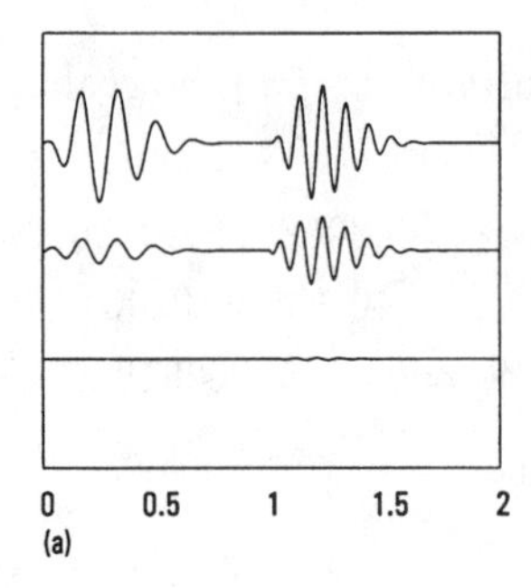

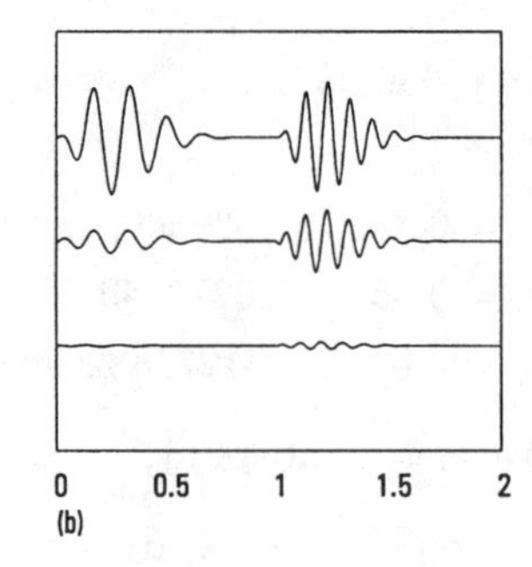

FIGURE 2.5
(a) Top: Signal. Middle: 1-level Daub4 fluctuation subsignal (multiplied by 1000 for comparison with the signal). Bottom: 1-level Daub6 fluctuation subsignal (also multiplied by 1000). (b) Similar graphs for 3-level Daub4 and Daub6 fluctuation subsignals (multiplied by 30).

Because of this property, the Daub6 transform will often produce smaller size fluctuation values than those produced by the Daub4 transform. The types of signals for which this occurs are the ones that are obtained from samples of analog signals that are at least three times continuously differentiable (at least over large portions of the analog signal). These kinds of signals are better approximated, over a large proportion of their values, by quadratic approximations rather than just linear approximations. Quadratic functions have curved graphs and can thereby provide superior approximations to the parts of the signal that are near to the turning points in its graph. To illustrate these ideas, consider the signal graphed at the top of Figure 2.5(a) and its 1-level Daub4 and Daub6 fluctuation subsignals graphed in the middle and at the bottom of the figure, respectively. This figure makes it clear that the Daub4 fluctuation values are significantly larger in magnitude than the Daub6 fluctuation values. It also shows that the largest magnitude Daub4 fluctuation values occur near the turning points in the graph of the signal. Similar graphs in Figure 2.5(b) illustrate the same ideas for the 3-level Daub4 and Daub6 fluctuation values.

When the goal is compression of signals, such as musical tones which often have graphs like the signal at the top of Figure 2.5(a), then the Daub6 transform can generally perform better at compressing the signal than the Daub4 transform. This is due to the larger number of Daub6 fluctuation values which can be ignored as insignificant. When the goal, however, is identifying features of the signal that are related to turning points in its graph, then the Daub4 transform can identify the location of these turning points more clearly as shown in Figure 2.5.

The other Daubechies wavelet transforms, the Daub$J$ transforms for $J = 8, 10, \ldots, 20$, are defined in essentially the same way. The scaling numbers

$\alpha_1, \ldots, \alpha_J$ satisfy

$$\alpha_1^2 + \alpha_2^2 + \cdots + \alpha_J^2 = 1, \tag{2.25a}$$

$$\alpha_1 + \alpha_2 + \cdots + \alpha_J = \sqrt{2}. \tag{2.25b}$$

And the wavelet numbers $\beta_1, \ldots, \beta_J$ are defined by

$$\beta_1 = \alpha_J,\ \beta_2 = -\alpha_{J-1},\ \beta_3 = \alpha_{J-2}, \ldots,\ \beta_{J-1} = \alpha_2,\ \beta_J = -\alpha_1. \tag{2.26}$$

These wavelet numbers satisfy the following identities (we set $0^0 = 0$ to enable a single statement):

$$0^L\beta_1 + 1^L\beta_2 + \ldots + (J-1)^L\beta_J = 0, \quad \text{for } L = 0, 1, \ldots, J/2 - 1. \tag{2.27}$$

These identities, along with (2.25b), imply the following property which is a generalization of Properties I and II above.

Property III. *If f is (approximately) equal to a polynomial of degree less than $J/2$ over the support of a k-level DaubJ wavelet W_m^k, then the k-level fluctuation value $\mathbf{f} \cdot \mathbf{W}_m^k$ is (approximately) zero.*

As with Property II above, this property implies that the Daub$J$ transform will produce a large number of small fluctuation values for a signal that is sampled from a smooth, many times continuously differentiable, signal. To put it another way, we can more closely approximate (obtain a better fit to) a wider range of signals if we can use higher degree polynomials with degree less than $J/2$, and yet still expect that the Daub$J$ transform will produce large numbers of small fluctuation values. As we showed in the previous chapter, when a wavelet transform produces a large number of small fluctuation values then we can obtain very effective compression and good noise removal.

One advantage of using a Daub$J$ wavelet with a larger value for J, say $J = 20$, is that there is an improvement in the resulting MRA for smoother signals (signals sampled from analog signals having more differentiability). For example, in Figure 2.6 we show the Daub20 MRA for the same signal analyzed previously with Haar and Daub4 wavelets. Notice that the Daub20 MRA is superior to both of these previous multiresolution analyses, especially for the lower resolution averaged signals.

We do not mean to suggest, however, that Daub20 wavelets are always the best. For example, for Signal 1 shown in Figure 1.4(a), the Haar wavelets do the best job of compression and noise removal (for reasons discussed in the previous chapter). As a simple comparison of the Haar, Daub4, and Daub20 wavelets, in Table 2.1 we list the minimum number of transform values needed to capture 99.99% of the energy in Signal 1. This table shows that the Haar transform does the best job, that the Daub4 transform is worse by a factor of 2, and that the Daub20 transform is the worst of all.

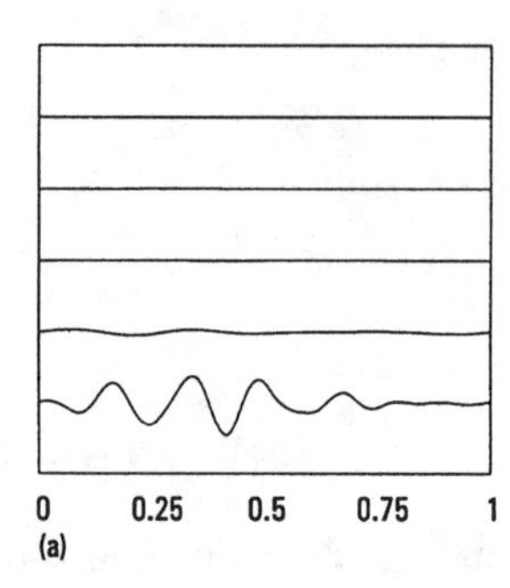

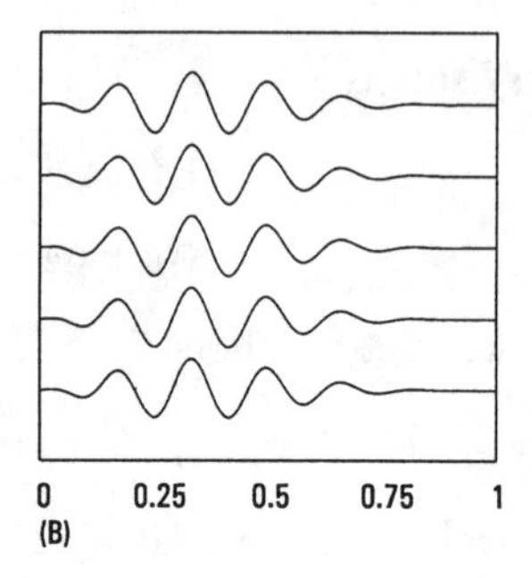

FIGURE 2.6
Daub20 MRA of the signal shown in Figure 1.1(a). The graphs are of the 10 averaged signals $\mathbf{A}^{10}$ through $\mathbf{A}^1$. Beginning with $\mathbf{A}^{10}$ on the top left down to $\mathbf{A}^6$ on the bottom left, then $\mathbf{A}^5$ on the top right down to $\mathbf{A}^1$ on the bottom right. Compare with Figures 1.3 and 2.3.

Table 2.1 Comparison of wavelet transforms of Signal 1

Wavelet transform	*Values for* 99.99% *energy*
Haar	51
Daub4	103
Daub20	205

The problem with Daub$J$ wavelets in terms of this signal is that these wavelets have longer supports than the Haar wavelets, all of their supports being at least twice as long, or longer, than the Haar wavelets. The Daub20 wavelets have the longest supports, with 1-level wavelets having supports of 20 time-units, and 2-level wavelets having supports of 58 time-units, and so on. Consequently, the percentage of Daub20 fluctuation values of this signal with significant energy will be high, due to the large number of Daub20 wavelets whose supports contain a point where a big jump in the signal's values occurs. A big jump in the signal's values induces corresponding jumps in the values of the scalar products that define the fluctuations, thus producing fluctuation values with significant energy.

Coiflets

We now turn to the description of another class of wavelets, the Coif$I$ wavelets. These wavelets are designed for the purpose of maintaining a close match between the trend values and the original signal values. Following a suggestion of Coifman, these wavelets were first constructed by Daubechies, who called them "coiflets." All of the Coif*I* wavelets are defined in a similar way; so we shall concentrate on the simplest case of Coif6 wavelets. The scaling numbers for the Coif6 scaling signals are listed in Table 2.2.

Table 2.2 Coif6 scaling numbers

$\alpha_1 = \frac{1-\sqrt{7}}{16\sqrt{2}}$,	$\alpha_2 = \frac{5+\sqrt{7}}{16\sqrt{2}}$,	$\alpha_3 = \frac{14+2\sqrt{7}}{16\sqrt{2}}$,
$\alpha_4 = \frac{14-2\sqrt{7}}{16\sqrt{2}}$,	$\alpha_5 = \frac{1-\sqrt{7}}{16\sqrt{2}}$,	$\alpha_6 = \frac{-3+\sqrt{7}}{16\sqrt{2}}$.

Using these scaling numbers, the first-level Coif6 scaling signals are defined by

$$\begin{aligned}
\mathbf{V}_1^1 &= (\alpha_3, \alpha_4, \alpha_5, \alpha_6, 0, 0, \ldots, 0, \alpha_1, \alpha_2) \\
\mathbf{V}_2^1 &= (\alpha_1, \alpha_2, \alpha_3, \alpha_4, \alpha_5, \alpha_6, 0, 0, \ldots, 0) \\
\mathbf{V}_3^1 &= (0, 0, \alpha_1, \alpha_2, \alpha_3, \alpha_4, \alpha_5, \alpha_6, 0, 0, \ldots, 0) \\
&\vdots \\
\mathbf{V}_{N/2}^1 &= (\alpha_5, \alpha_6, 0, 0, \ldots, 0, \alpha_1, \alpha_2, \alpha_3, \alpha_4)
\end{aligned} \tag{2.28}$$

Notice that there are wrap-arounds for $\mathbf{V}_1^1$ and $\mathbf{V}_{N/2}^1$.

The Coif6 wavelet numbers are defined by

$$\beta_1 = \alpha_6,\ \beta_2 = -\alpha_5,\ \beta_3 = \alpha_4,\ \beta_4 = -\alpha_3,\ \beta_5 = \alpha_2,\ \beta_1 = -\alpha_1 \tag{2.29}$$

and these wavelet numbers determine the first-level Coif6 wavelets as follows:

$$\begin{aligned}
\mathbf{W}_1^1 &= (\beta_3, \beta_4, \beta_5, \beta_6, 0, 0, \ldots, 0, \beta_1, \beta_2) \\
\mathbf{W}_2^1 &= (\beta_1, \beta_2, \beta_3, \beta_4, \beta_5, \beta_6, 0, 0, \ldots, 0) \\
\mathbf{W}_3^1 &= (0, 0, \beta_1, \beta_2, \beta_3, \beta_4, \beta_5, \beta_6, 0, 0, \ldots, 0) \\
&\vdots \\
\mathbf{W}_{N/2}^1 &= (\beta_5, \beta_6, 0, 0, \ldots, 0, \beta_1, \beta_2, \beta_3, \beta_4)
\end{aligned} \tag{2.30}$$

As with the Coif6 scaling signals, there are wrap-arounds for the first and last wavelets.

The Coif6 scaling numbers satisfy the following identity

$$\alpha_1^2 + \alpha_2^2 + \alpha_3^2 + \alpha_4^2 + \alpha_5^2 + \alpha_6^2 = 1 \tag{2.31}$$

which implies that each Coif6 scaling signal has energy 1. Because of (2.29), it follows that each Coif6 wavelet also has energy 1. Furthermore, the wavelet numbers satisfy

$$\beta_1 + \beta_2 + \beta_3 + \beta_4 + \beta_5 + \beta_6 = 0, \tag{2.32a}$$

$$0\beta_1 + 1\beta_2 + 2\beta_3 + 3\beta_4 + 4\beta_5 + 5\beta_6 = 0. \tag{2.32b}$$

These equations show that a Coif6 wavelet is similar to a Daub4 wavelet in that it will produce a zero fluctuation value whenever a signal is linear over its support. The difference between a Coif6 wavelet and a Daub4 wavelet lies in the properties of the scaling numbers. The Coif6 scaling numbers satisfy

$$\alpha_1 + \alpha_2 + \alpha_3 + \alpha_4 + \alpha_5 + \alpha_6 = \sqrt{2}, \tag{2.33a}$$

$$-2\alpha_1 - 1\alpha_2 + 0\alpha_3 + 1\alpha_4 + 2\alpha_5 + 3\alpha_6 = 0, \tag{2.33b}$$

$$(-2)^2\alpha_1 + (-1)^2\alpha_2 + 0^2\alpha_3 + 1^2\alpha_4 + 2^2\alpha_5 + 3^2\alpha_6 = 0. \tag{2.33c}$$

Equation (2.33a) implies, as usual, that Coif6 trend values are averages of successive values of a signal $\mathbf{f}$ (with wrap-around when $\mathbf{f} \cdot \mathbf{V}_1^1$ and $\mathbf{f} \cdot \mathbf{V}_{N/2}^1$ are computed). The second two equations, however, are entirely new. No Daub$J$ scaling numbers satisfy any equations of this type. These three equations have an important consequence. *When a signal consists of sample values of an analog signal, then a Coif6 transform produces a much closer match between trend subsignals and the original signal values than can be obtained with any of the DaubJ transforms.* By a close match between trends and signal values, we mean that the following approximations hold to a high degree of accuracy:

$$\mathbf{a}_m^1 \approx \sqrt{2}g(t_{2m}), \qquad \mathbf{a}_m^2 \approx 2g(t_{4m}) \tag{2.34}$$

Similar approximations will hold for higher levels, but the accuracy generally decreases as the number of levels increases.

As an example of (2.34), consider the signal graphed in Figure 2.2(a). This signal is obtained from 2^{14} sample values of the function

$$g(x) = 20x^2(1-x)^4 \cos 12\pi x \tag{2.35}$$

over the interval $[0, 1)$. When a 2-level Daub4 transform is performed on this signal then we obtain the graph shown in Figure 2.2(b). A 2-level Coif6 transform looks much the same. Computing the maximum error between the 2-level Daub4 trend values and samples of $2g(4x)$ over the interval $[0, .25)$, we obtain 3.76×10^{-3}. The maximum error in the Coif6 case is 4.84×10^{-7}, which is much smaller. For the 1-level transforms we find that the maximum error between the first trend and samples of $\sqrt{2}g(2x)$ is 8.87×10^{-4} in the Daub4 case, and 8.59×10^{-8} in the Coif6 case. This property of trends providing close approximations of the analog signal, which is shared by all the Coif$I$ transforms, provides a useful means of interpreting the trend subsignals.

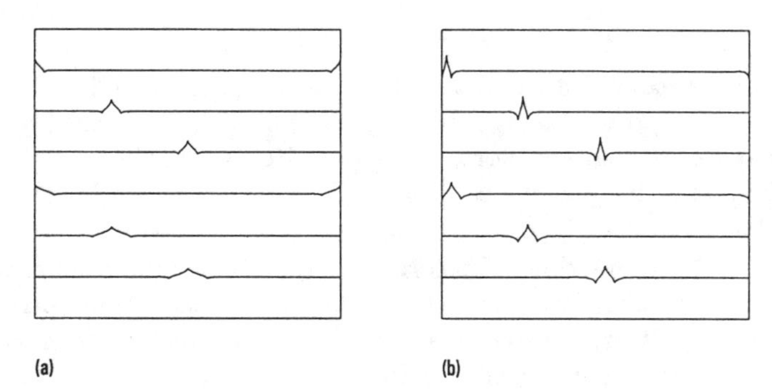

FIGURE 2.7
(a) The top 3 signals are 5-level Coif6 scaling signals, V^5_1, V^5_8, and V^5_{16}. The bottom 3 signals are 6-level scaling signals V^6_1, V^6_4, and V^6_8. (b) The top 3 signals are 5-level Coif6 wavelets, W^5_1, W^5_8, and W^5_{16}. The bottom 3 signals are 6-level wavelets W^6_1, W^6_4, and W^6_8.

Another interesting feature of Coif*I* scaling signals and wavelets is that their graphs are nearly symmetric. For example, in Figure 2.7 we graph several Coif6 scaling signals and wavelets. Notice how these Coif6 signals (especially if we discount the wrap-around effects) are much closer to being symmetric than the Daub4 signals graphed in Figure 2.1.

2.4 Compression of audio signals

One of the fundamental applications of wavelet transforms is the compression of signals. We outlined a basic method for wavelet transform compression in Section 1.5. In that section we focused on the Haar wavelet transform; in this section we shall work with the Daubechies wavelet transforms. We shall also discuss the problem of *quantization,* which we omitted from our first treatment of compression.

Recall that the basic method of wavelet transform compression consisted of setting equal to zero all transform values whose magnitudes lie below a threshold value. The compressed version of the signal consists of the significant, non-zero, values of the transform which survived the thresholding, along with a significance map indicating their indices. Decompression consists in using the significance map and the significant transform values to reconstruct the thresholded transform, and then performing an inverse wavelet transform to produce an approximation of the original signal. Compression works well when very few, high-energy, transform values capture most of the energy of the signal. For instance, consider again the two Signals 1 and 2 examined in Section 1.5. We saw that Signal 1 can be very effectively compressed using a Haar transform. This is revealed by the Energy map for its Haar transform in Figure 1.4(c), which shows that the Haar transform effectively captures most of the energy of Signal 1 in relatively few values.

None of the Daubechies transforms can do a better job compressing Signal 1. We have already examined why the Haar transform performs so well on Signal 1. In fact, the Haar transform and a related transform called the Walsh transform[3] have been used for many years as tools for compressing *piecewise constant* signals like Signal 1. We also saw, however, that Signal 2 does not compress particularly well using the Haar transform. This is explained by an examination of its Energy map shown in Figure 1.5(c). Let's instead try compressing Signal 2 using one of the Daubechies transforms. Signal 2 consists of $4096 = 2^{12}$ points; so we will use a 12-level transform, say a Coif30 transform. In Figure 2.8, we show the results of applying a Coif30 wavelet transform compression on Signal 2. It is interesting to compare this figure with Figure 1.4. It is clear that the 12-level Coif30 transform compresses Signal 2 just as well as the Haar transform compresses Signal 1. In fact, by using only the top 125 highest magnitude Coif30 transform values—which can be done by choosing a threshold of .00425—the compressed signal captures 99.99% of the energy of Signal 2. This compressed signal is shown in Figure 2.8(d). Since $4096/125 \approx 32$, the compressed signal achieves a 32:1 compression ratio. Here we are ignoring issues such as quantization and compression of the significance map. We now turn to a brief discussion of these deeper issues of compression; this initial treatment will be expanded upon in the next section.

Quantizing signal values, compressing the significance map

A digital audio signal typically consists of integer values that specify *volume levels.* The two most common ranges of volume levels are either $256 = 2^8$ volume levels, which require 8 bits to describe, or $65536 = 2^{16}$ volume levels, which require 16 bits to describe. An analog audio signal is *quantized* by a mapping from the recorded volume level to one of these two ranges of volume levels. Therefore, a discrete audio signal of length N will be initially described by either $8N$ or $16N$ bits, depending on which range of volume levels is used for recording. The 8-bit range is frequently used for voices in telephone transmission, where high fidelity is sacrificed for speed of transmission in order to accommodate the large number of signals that must be transmitted. The 16-bit range is frequently used for music, where high fidelity is most valued.

[3]The Walsh transform is described in Section 4.1.

Table 2.3 Encoding 16 volume levels using 4 bits

Volume level	*Encoding*
−24	1111
−21	1110
⋮	⋮
−1	1000
0	0000
1	0001
⋮	⋮
18	0110
21	0111

The most commonly employed quantization method for sampled analog signals is *uniform scalar quantization.* This method simply divides the range of volume levels into a fixed number of uniform width subintervals and rounds each volume level into the midpoint of the subinterval in which it lies. For instance, in Figure 2.9(a), we show a simple uniform scalar quantization map that encodes volume levels using 4 bits. The volume interval $[-24, 21]$ is divided into $16 = 2^4$ equal width subintervals, with all volumes that are below -24 being truncated to -24 and all volumes that are greater than 21 being truncated to 21. These 16 volume levels can be encoded as shown in Table 2.3. The asymmetry in this uniform quantization decreases as more subintervals, i.e., more bits, are used.

In order to take into account the number of bits used per point (bpp), which is either 8 bpp or 16 bpp in a quantized audio signal, we must also quantize the transform coefficients. That is, we must use only a finite number of bits to describe each transform value, and the number of bpp must be significantly less for the compressed signal than for the original signal.

As an initial example of handling quantization, consider again Signal 2 shown in Figure 2.8(a). This signal was generated from 4096 uniform samples of an analog signal. If this signal is uniformly scalar quantized with 16 bpp and played as an audio signal at a rate of 8820 samples per second,[4] multiplying its volume by a factor of 32000, then the resulting sound resembles two low notes played on a clarinet. The 16-bit quantized version of the signal has a graph that is almost identical to the one shown in Figure 2.8(a). If a Coif30 transform is performed on this quantized signal,

[4]Volume levels are sent to the sound system at a rate of 8820 values per second.

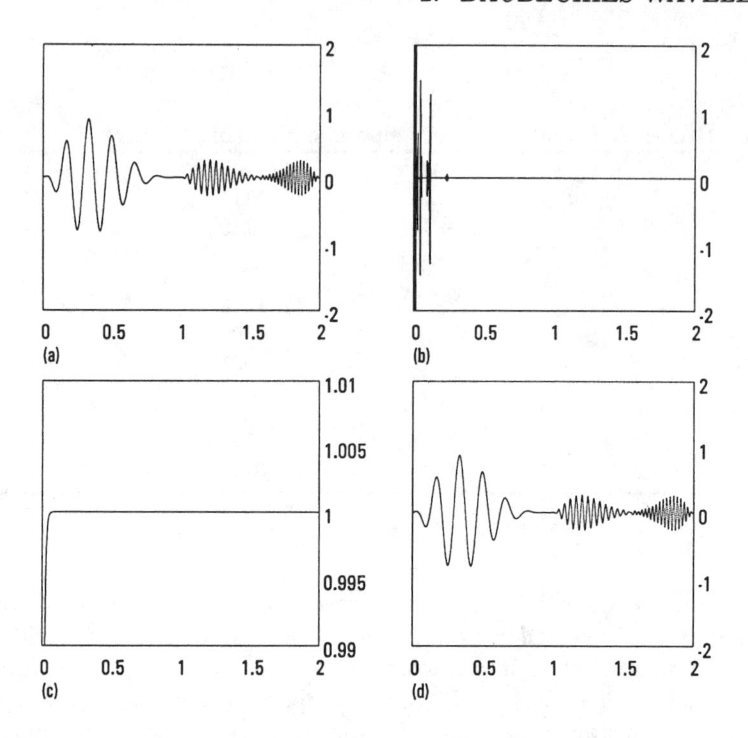

FIGURE 2.8
(a) Signal 2, 4096 values. (b) 12-level Coif30 transform. (c) Energy map of the Coif30 transform. (d) 32:1 compression of Signal 2, 99.99% of energy of Signal 2.

then a transform is produced which is virtually indistinguishable from the graph shown in Figure 2.8(b). To quantize this transform, we proceed as follows. The quantization map used is similar to the one shown in Figure 2.9(b). This is called *uniform quantization with a dead-zone.* The values in the subinterval $(-T, T)$ are the insignificant values whose magnitudes lie below a threshold value of T. Since these values will not be transmitted they are not encoded by the quantization. The remainder of the range of transform values lies in the two intervals $[-M, -T]$ and $[T, M]$, where M is the maximum for all the magnitudes of the transform values. These two intervals are divided into uniform width subintervals and each transform value is rounded into the midpoint of the subinterval containing it. For Signal 2, the value of M is 9.4, and a threshold value of $T = 9.4/2^7$ results in only 100 significant values. These significant values can then be encoded using 8 bits, 7 bits for the levels of magnitude and 1 bit for signs. As can be seen from Figure 2.8(b), the significant values of the transform lie in the interval $[0, .25)$. In fact, the significant values of the quantized transform lie among the first 256 values. Consequently the bits of value 1

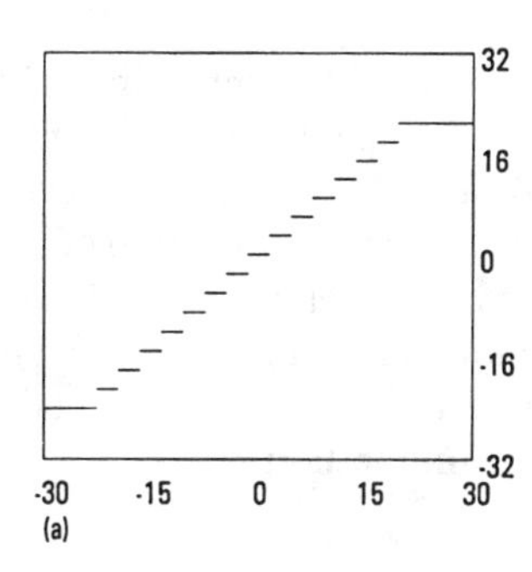

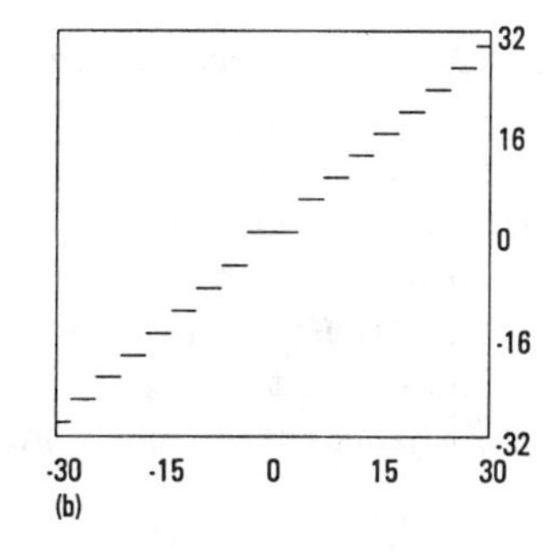

FIGURE 2.9
(a) Uniform, scalar quantization. (b) Uniform quantization, with a dead-zone containing the 0-level. The quantized values outside the dead-zone can be encoded using 4 bits.

in the significance map lie only within the first 256 bits. Therefore, this significance map can be transmitted in a very compressed form by just transmitting the first 256 bits, letting the decompression program supply the remaining bits, which are all 0. The result is that

$$\frac{256 + 8 \cdot 100}{4096} \approx 0.27 \text{ bpp}$$

are needed to transmit the compressed signal. This represents a compression ratio of 60:1. Even more important, when the decompressed signal is played, the resulting sound is indistinguishable from the original signal's sound. This is known as *perceptually lossless* compression.

This example was meant to illustrate the basic idea of quantization and its relation to wavelet transform compression of audio signals. In the next section we shall delve further into some of the fascinating complexities of quantization and compression of signals.

2.5 Quantization, entropy, and compression

In this section we shall examine some of the deeper questions involved with quantization and compression. One of the interesting features of quantizing wavelet transforms is the connection between structural properties of these transforms and basic ideas from information theory, such as entropy. This connection can be exploited to improve upon the basic compression results described in the preceding section.

Let's begin with a specific signal to be compressed, a recording of the author speaking the word *greasy*. A graph of the intensity values of this recording is shown in Figure 2.10(a). Notice that this graph consists of three main sections, summarized in Table 2.4. This signal exhibits within a short time-span a number of different linguistic effects, such as a quick transition from a low pitch *gr* sound to a higher pitch *e* sound, and then another quick transition to a very chaotically oscillating *s* sound, with a final transition back to a high pitch *y* sound. These effects make *greasy* a challenging test signal for any compression method.

Table 2.4 Main sections of *greasy*

Section	*Time Interval*	*Sound*
1	$[0.00, 0.20]$	grea
2	$[0.25, 0.35]$	s
3	$[0.40, 0.60]$	y

The intensity values for this recording of *greasy* were quantized—using an 8-bit scalar quantization—as a part of the recording process. The sequences of bits encoding these 256 intensity values look like 01000001 or 11001000, and so on. A first bit of 1 indicates a negative intensity, while a first bit of 0 indicates a positive intensity. These bit sequences correspond to the integers $k = 0$ to $k = 255$, which we shall refer to as the *intensity levels.* To some extent this use of equal length bit sequences for all intensity levels is wasteful. This is indicated by the histogram of frequencies of occurrence of each intensity level shown in Figure 2.10(b). Since the most commonly occurring intensity level is the zero level, we can save bits if we encode this level with a single bit, such as 0.

By using shorter length bit sequences for the most commonly occurring intensity levels, and longer sequences for less commonly occurring intensity levels, we can reduce the total number of bits used. The idea is similar to Morse code where, for instance, the commonly occurring English letters a and e are encoded by the short sequences of dots and dashes $\cdot -$ and $\cdot$, respectively, while the less commonly occurring English letters q and v are encoded by the longer sequences $- - \cdot -$ and $\cdot \cdot \cdot -$, respectively. This procedure is made mathematically precise by fundamental results from the field known as *information theory.*

It is beyond the scope of this primer to provide a rigorous treatment of information theory. We shall just outline the basic ideas, and show how they apply to compressing *greasy*. Suppose that $\{p_k\}$ are the relative frequencies of occurrence of the intensity levels $k = 0$ through $k = 255$; that is, each p_k is an ordinate value in the histogram in Figure 2.10(b). Thus $p_k \geq 0$ for each k and $p_0 + p_1 + \cdots + p_{255} = 1$. These facts make it tempting to interpret each number p_k as a probability for the occurrence of k. Although these numbers p_k are not probabilities, nevertheless, a deterministic law governing the production of the intensity levels in *greasy* is *a priori* unknown to us. In fact, section 2 of *greasy,* as an isolated sound, is very similar to the random static background noise considered in the next section. There are deterministic models for producing sections 1 and 3, involving combinations of sinusoidal signals, but these are *a posteriori* models based on the recorded sound itself. In any case, let's see what consequences follow from treating the numbers p_k as probabilities for the occurrence of the intensity levels k. Let $\mathcal{L}_k$ be the length of each bit sequence that is used to encode the intensity level k. Then the *average length* $\overline{\mathcal{L}}$ of a lossless encoding of the k's is defined to be

$$\overline{\mathcal{L}} = p_0\mathcal{L}_0 + p_1\mathcal{L}_1 + \cdots + p_{255}\mathcal{L}_{255} \,. \tag{2.36}$$

The famous Shannon Coding Theorem tells us that $\overline{\mathcal{L}}$ satisfies the following inequality [if $p_k = 0$, then $p_k \log_2(1/p_k)$ is set equal to 0]:

$$\overline{\mathcal{L}} \geq p_0 \log_2 \frac{1}{p_0} + p_1 \log_2 \frac{1}{p_1} + \cdots + p_{255} \log_2 \frac{1}{p_{255}} \,. \tag{2.37}$$

The quantity on the right side of (2.37) is called the *entropy* for the probabilities p_k.

Inequality (2.37) says that the average length of any lossless encoding of the intensity levels k cannot be less than the entropy of the probabilities p_k in the histogram for these intensity levels. Or another, more accurate, way of putting things is that *a lossless encoding technique, such as Huffman coding or arithmetic coding, which is based on the relative frequencies p_k in the histogram for the k's, cannot achieve an average length less than the entropy.* For obvious reasons, these lossless coding techniques are called *entropy codings.*

For *greasy,* the entropy is found to be 5.43. Therefore, it is impossible to make an entropy coding of the intensity levels for *greasy* with any set of bit sequences whose average length is less than 5.43 bits. It should also be noted that, using either Huffman coding or arithmetic coding, it is possible to get within 1 bit or less of the entropy. For reasons of space, we shall not describe the methodologies of these encoding techniques; for another matter, these techniques are very well-known and there are many excellent descriptions of them to be found in the references for this chapter. Huffman codes are guaranteed, by the way in which they are constructed, to always achieve an average length that is within 1 bit of the entropy; while arithmetic codes can get asymptotically close to the entropy as the number of values to be encoded increases. Therefore, as a rough estimator of the average length of a well-chosen lossless encoding of the intensity levels we shall add 0.5 to the entropy. Using this estimator, we find that the 16,384

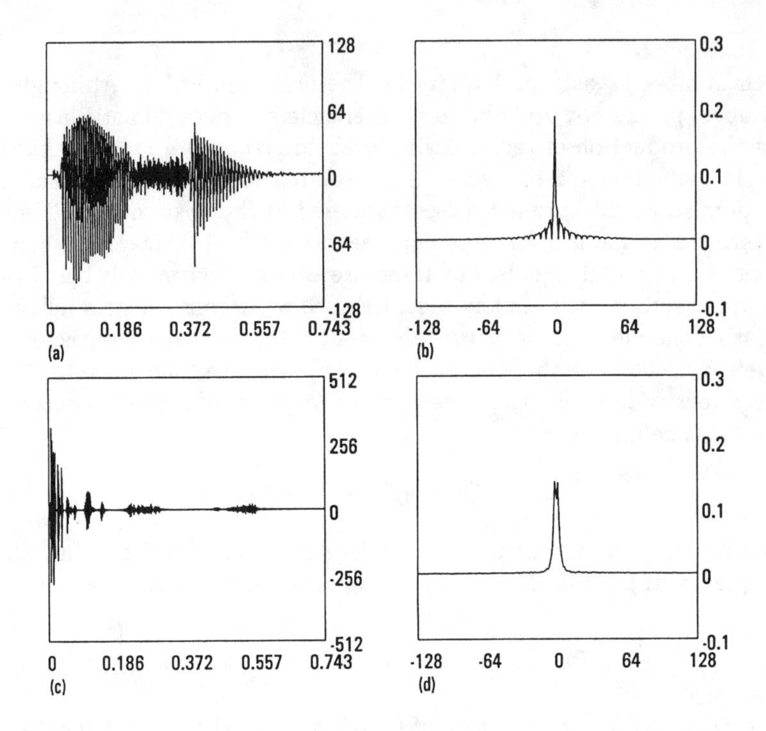

FIGURE 2.10
(a) The signal, *greasy.* (b) Histogram of 8-bit quantization of intensity levels of *greasy.* (c) 14-level Coif30 transform of *greasy.* (d) Histogram of 8-bit dead-zone quantization of the transform values.

points of *greasy* can be expected to be entropy encoded with a total of $16,384 \times 5.93$ bits, i.e., using about $97,000$ bits. This is not a particularly effective compression, since it still represents 5.93 bpp versus 8 bpp for the original signal.

The basis of wavelet transform encoding, as we explained in the last section, is to allow some inaccuracy resulting from quantizing transform values in order to achieve greater compression than by lossless methods. For instance, in Figure 2.10(c), we show a 14-level Coif30 transform of the *greasy* recording, and in Figure 2.10(d) we show a histogram of an 8-bit dead-zone quantization of this transform. Comparing the two histograms in Figure 2.10, we see that the histogram for the dead-zone quantization of the transform values is more narrowly concentrated than the histogram for the scalar quantization of the signal values. In fact, the entropy for the histogram in Figure 2.10(d) is 4.34, which is smaller than the entropy 5.43 for the histogram in Figure 2.10(c). Using our estimator for average coding length, we estimate that a lossless encoding of the quantized transform values will have an average length of 4.84. Consequently, the 3922 non-zero

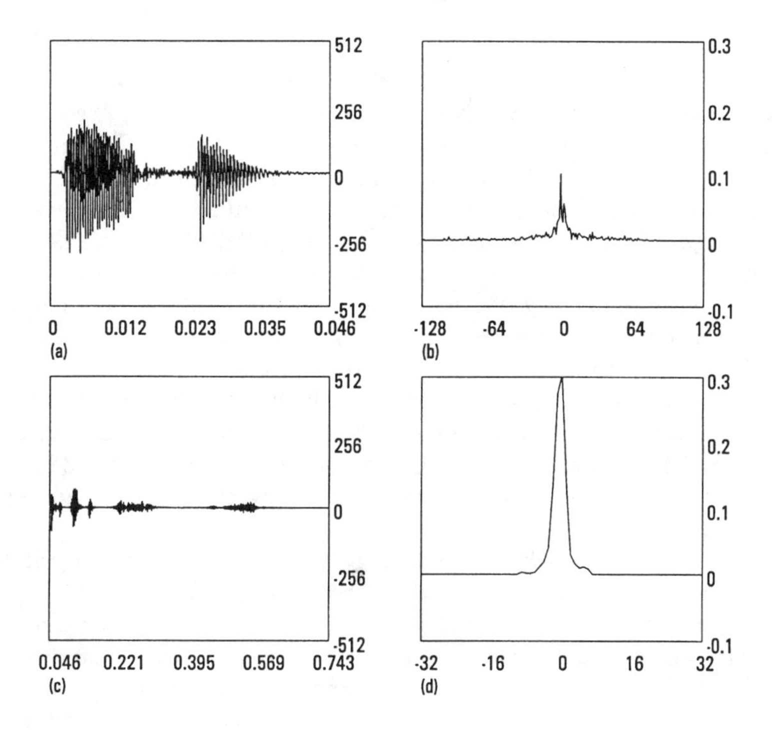

FIGURE 2.11
(a) Fourth trend of 4-level Coif30 transform of *greasy*. (b) Histogram of 8-bit dead-zone quantization of fourth trend. (c) Fluctuations of 4-level Coif30 transform. (d) Histogram of 6-bit dead-zone quantization of fluctuations.

quantized transform values can be expected to be losslessly encoded with about $3922 \times 4.84 \approx 18,922$ bits. This is a significant reduction in the number of bits; even if absolutely no compression was done on the 16,384 bits in the significance map, there would still be a large improvement in compression over a lossless compression of *greasy*. It is very important to note that—even though there is some error introduced by the quantization of the transform values—when an inverse transform is performed, the resulting signal is extremely difficult to distinguish from the original when played over a computer sound system.

As another example of the possibilities available with wavelet transform compression, we shall examine a slightly different approach to compressing *greasy*. In Figures 2.11(a) and 2.11(c), we show a 4-level Coif30 transform of *greasy*. The fourth trend is graphed in Figure 2.11(a) and the 4 fluctuations are graphed in Figure 2.11(c). An 8-bit dead-zone quantization of the fourth trend has the histogram shown in Figure 2.11(b), with an entropy of 6.19. There are 793 non-zero coefficients; so encoding requires about

$793 \times 6.69 \approx 5305$ bits. To quantize the fluctuation values we chose to use only 6 bits, not 8 bits as we used for the 14-level transform. A comparison of Figures 2.10(c) and 2.11(c) makes it apparent why we did this. More intensity levels are needed with the 14-level transform in order to ensure accuracy because of the many larger fluctuation values at higher levels [due to the higher peaks on the left side of Figure 2.10(c)]. This 6-bit dead-zone quantization of the four fluctuations has the histogram shown in Figure 2.11(d), with an entropy of 2.68. Since there are 2892 non-zero quantized values, an encoding requires about $2892 \times 3.18 \approx 9197$ bits. The total number of bits estimated for this 4-level transform compression is about $14,502$ bits. This compares favorably with the $18,982$ bits estimated for the 14-level transform compression. The improvement in compression for the 4-level transform is due to the decrease in entropy from 4.34 to 2.68 for the first four fluctuations brought about by the change made in the quantization.

This last example only begins to suggest some of the many possibilities available for adaptation of the basic wavelet transform compression procedure. For instance, one possibility is to compute an entropy for a separate quantization of each fluctuation and separately encode these quantized fluctuations. This does generally produce improvements in compression. For example, suppose a 4-level Coif30 transform of *greasy* is quantized using 8 bpp for the trend and 6 bpp for the four fluctuations, and separate entropies are calculated for the trend and for each of the four fluctuations. Then the estimated total number of bits needed is $11,305$. This is an improvement over the $14,502$ bits previously estimated.

2.6 Denoising audio signals

As we saw in Section 1.6, the problems of compression and noise removal are closely related. If a wavelet transform can effectively capture the energy of a signal in a few high-energy transform values, then additive noise can be effectively removed as well. We introduced a basic method, called *thresholding,* for removing noise, and illustrated this method using the Haar transform. Now, in this section, we shall threshold the Daubechies wavelet transforms. Besides discussing a simple illustrative example, we also shall give some justification for why thresholding works well for the random noise often encountered in signal transmission, and provide an example of denoising when the noise is a combination of pop noise and random noise.

Let's quickly review the basic steps for removing additive noise using the Threshold Method. A threshold value T is chosen for which all transform values that are lesser in magnitude than T are set equal to zero. By performing an inverse wavelet transform on the thresholded transform, an estimation of the original uncontaminated signal is obtained. In Chapter 1 we saw that the Haar transform was able to effectively remove the noise from Signal A, as shown in Figure 1.6. Signal A was created by adding random noise to Signal 1, whose graph is shown in Figure 1.4. Because of the connection between compression and noise removal, and because the Haar transform is the most effective transform for compressing Signal 1, it follows that the Haar transform is also the most effective transform for denoising Signal A.

With the other noisy signal examined in Section 1.6, Signal B, we found that the Haar transform did a rather poor job of denoising. The explanation for this lies again in the connection between compression and threshold denoising. Signal B was created by adding random noise to Signal 2, but the Haar transform is not an effective tool for compressing Signal 2; it produces too many low magnitude transform values which are obscured by the noise. We saw, however, in Section 2.4 that the Coif30 transform is a very effective tool for compressing Signal 2. Therefore, let's apply it to denoising Signal B.

In Figure 2.12 we show the basic steps in a threshold denoising of Signal B using a 12-level Coif30 transform. Comparing the 12-level Coif30 transforms of Signal B and Signal 2, we see that the addition of the noise has contributed a large number of small magnitude, low-energy values to the transform of Signal 2. Nevertheless, most of the high-energy values of the transform of Signal 2 are still plainly visible in the transform of Signal B, although their values have been altered slightly by the added noise. Therefore, we can eliminate the noise using thresholding, as indicated by the two horizontal lines in Figure 2.12(b). All transform values between ± 0.2 are set equal to zero, and this produces the thresholded transform shown in Figure 2.12(c). Comparing this thresholded transform with the transform of Signal 2, shown in Figure 2.8(b), we can see that thresholding has produced a fairly close match. The most noticeable difference is a loss of a small number of values located near 0.25, which had such small magnitudes that they were obscured by the addition of the noise. Since the two transforms are such a reasonably close match, it follows that the inverse transform of the thresholded transform produces a denoised signal that is a close match of Signal 2. In fact, the RMS Error between the denoised signal and Signal 2 is 0.14, which is a four-fold decrease in the RMS Error of 0.57 between Signal B and Signal 2.

Choosing a threshold value

One of the most attractive features of wavelet threshold denoising is that, for the type of random noise frequently encountered in signal transmission,

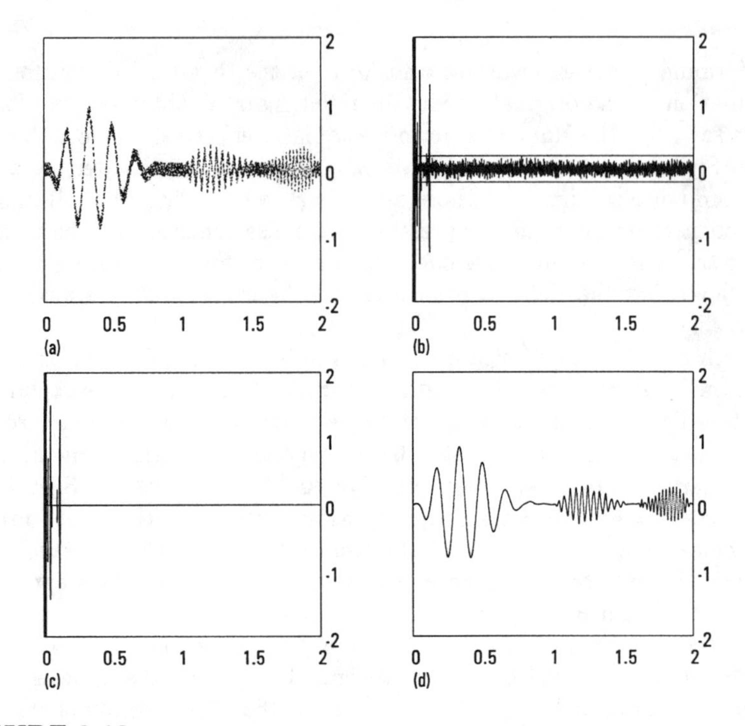

FIGURE 2.12
(a) Signal B. (b) 12-level Coif30 transform, with thresholding indicated by two horizontal lines at ±0.2. (c) Thresholded transform. (d) Denoised signal; compare with Figures 2.8(a) and 1.6(d).

it is possible to automatically choose a threshold for denoising *without any prior knowledge of the signal.* In Figure 2.13(a) we show an example of this type of random noise; it is called *Gaussian noise.* A technical definition of Gaussian random noise is beyond the scope of this book; in this subsection we shall only give an informal introduction to some of the main ideas.

If the noise shown in Figure 2.13(a) is played over a sound generator, it produces the familiar static sound that can be heard in noisy radio transmissions and analog recordings. A histogram of frequencies of occurrence of intensity levels for this noise is shown in Figure 2.13(b). The bell-shaped curve that this histogram approximates is an indication that this noise is Gaussian. Using elementary statistical formulas on the noise values, we can estimate the mean μ and standard deviation σ of the probability density function (the bell-shaped curve) that this histogram approximates. We find that the mean is approximately 0 and that the standard deviation is approximately 0.579. One of the consequences of the Conservation of Energy Property of the Daubechies wavelet transforms—more precisely, the

Table 2.5 Comparison of three noise histograms

Threshold	*% below, Theory*	*% below, 2.13(a)*	*% below, 2.13(c)*
σ	68.72	68.87	69.94
2σ	95.45	95.70	95.39
3σ	99.73	99.74	99.66
4σ	99.99	100.00	99.98

orthogonality of the matrix form for these transforms[5]—is that they preserve the Gaussian nature of the noise. For example, in Figure 2.13(c) we show a Coif30 wavelet transform of the random noise in Figure 2.13(a). Its histogram in Figure 2.13(d) approximates a bell-shaped curve, and the mean and standard deviation of the transformed noise are calculated to be approximately 0 and 0.579, the same values that we found for the original noise.

These facts imply that the transformed noise values will be similar in magnitude to the original noise values and, what is more important, that *a large percentage of these values will be smaller in magnitude than a threshold equal to a large enough multiple of the standard deviation* σ. To see this last point more clearly, consider the data shown in Table 2.5. In this table, the first column contains four thresholds which are multiples of the standard deviation $\sigma = 0.579$. The second column lists the percentages of values of random numbers having magnitudes that are less than these thresholds, assuming that these random numbers obey a Gaussian normal probability law with mean 0 and standard deviation σ.[6] The third and fourth columns contain the percentages of the magnitudes from the noise signals in Figure 2.13(a) and 2.13(c) which lie below the thresholds.

Based on the results of this table, we shall use the following formula:

$$T = 4.5\sigma \tag{2.38}$$

for setting the threshold value T. In our next example, we shall make use of (2.38) for choosing a threshold. The standard deviation σ can be estimated from a portion of the transform which consists largely of noise values. Generally this is the case with the first level fluctuation because the first level fluctuation values from the original signal are typically very small. When Formula (2.38) is used we expect that well over 99% of the

[5]Orthogonality of the Daubechies wavelet transforms was discussed in Section 2.2.

[6]That is, the probability of a number x having magnitude less than T is equal to the area under the curve $\frac{1}{\sigma\sqrt{2\pi}}e^{-x^2/2\sigma^2}$ from $-T$ to T.

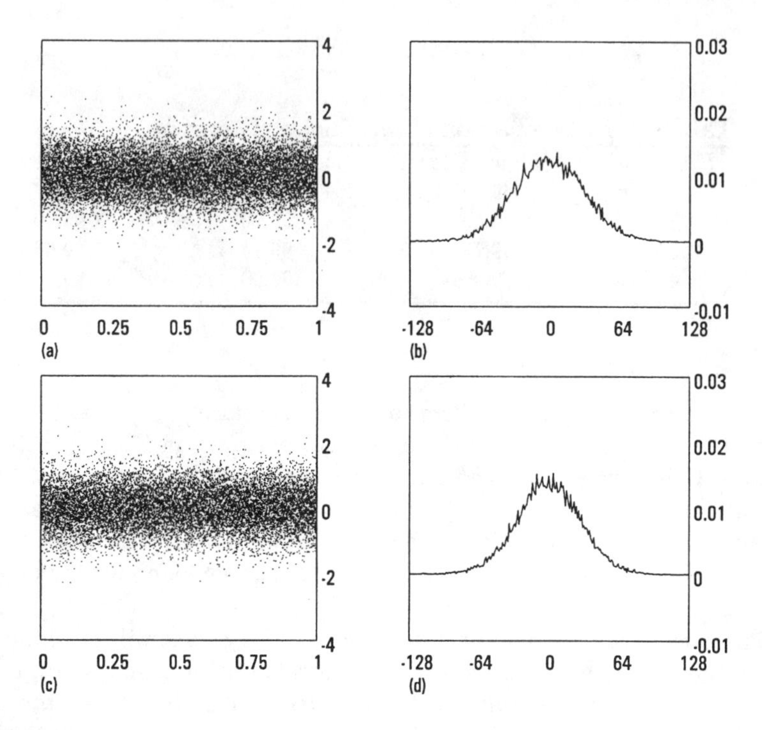

FIGURE 2.13
(a) Gaussian noise. (b) Histogram of the noise. (c) Coif30 transform of the noise. (d) Histogram of the transformed noise.

noise values will be removed from the transform, and this requires essentially no knowledge of the original, uncontaminated signal. Preventing this thresholding from removing too many transform values from the original signal, however, depends on how well the transform compresses the energy of the signal into a few high-magnitude values which stand out from the threshold. With the Daubechies transforms this will occur with signals that are sampled from analog signals that are smooth. This is because so many of the fluctuation values are small, and that leaves only a few high-magnitude transform values to account for the energy. Of course it is always possible—by adding noise with a sufficiently high standard deviation—to produce such a high threshold that the signal's transform values are completely wiped out by thresholding. Wavelet thresholding is powerful, but it cannot perform magic.

Removing pop noise and background static

We close this section by discussing another example of noise removal. In Figure 2.14(a) we show the graph of a noisy signal. The signal consists of

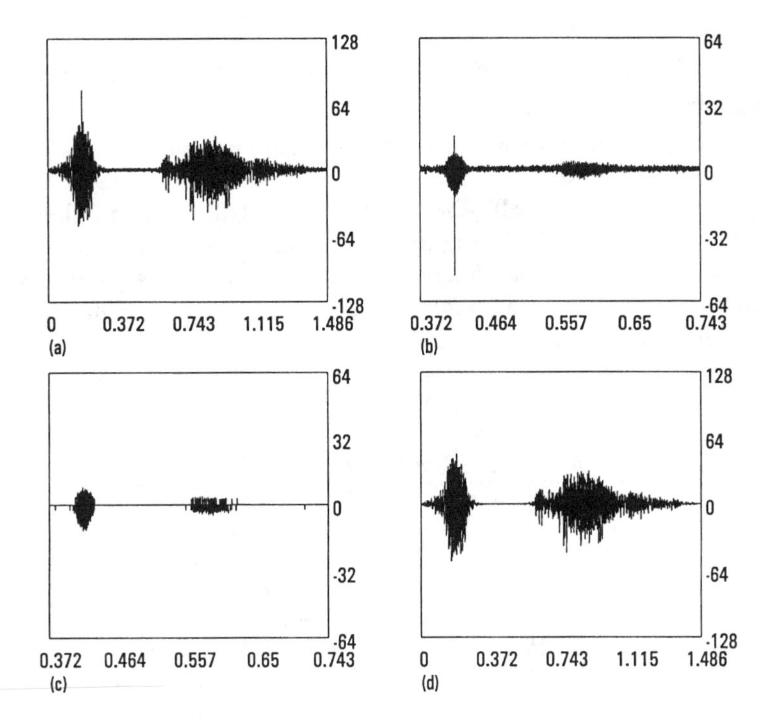

FIGURE 2.14
(a) Noisy whistle. (b) Second fluctuation of the 15-level Coif18 transform. (c) Second fluctuation of the denoised transform. (d) Denoised whistle.

Gaussian random noise and pop noise added to a recording of the author making a whistling sound. The pop noise consists of the spike located near 0.186, which is plainly visible in the graph. When this signal is played over a computer sound system, the whistle can be plainly heard but there is an annoying static background and a rather loud, short, popping sound near the beginning of the whistle. Using wavelet techniques we can remove essentially all of the random noise and greatly reduce the volume of the pop noise.

To remove the noise, we use a Coif18 wavelet transform of the signal. In Figure 2.14(b) we show the second fluctuation subsignal of a Coif18 transform of the noisy signal. The other fluctuation subsignals are similar in appearance; so by describing how to modify this particular fluctuation we shall be describing how to handle the other fluctuations as well. The random noise, or background static, appears as random oscillations in the wavelet transform as we noted above. In fact, on the interval $[.46, .51]$ the fluctuation values appear purely random; so we make an estimate of the standard deviation of the transform noise over this subinterval, obtaining

$\sigma \approx 0.617$. By Formula (2.38), we should set T equal to 2.7765. Rounding up slightly, we set $T = 2.78$.

In contrast to the random noise, the pop noise produces unusually large magnitude fluctuation values that stand out from the vast majority of the fluctuation values. We shall refer to these unusually large fluctuation values as *outliers*. For example, in Figure 2.14(b) there are two outliers that are clearly visible in the second fluctuation. They appear as two spikes on the left side of the figure. To remove the pop noise we must eliminate these outliers from the transform.

Table 2.6 Acceptance bands for fluctuations of noisy signal

Fluctuation	*Acceptance band*
1	$(-\infty, -T] \cup [T, \infty)$
2	$(-12.5, -T] \cup [T, 9)$
3	$(-\infty, -T] \cup [T, \infty)$
4	$(-67, -T] \cup [T, 67)$
5 to 15	$(-\infty, -T] \cup [T, \infty)$

To remove both the random noise values and the outliers from the pop noise, we set *acceptance bands* for each fluctuation subsignal. Values that lie in the acceptance band for each fluctuation are retained and all other values are set equal to zero. These acceptance bands, which were obtained by a visual inspection of the transform, are summarized in Table 2.6. The second fluctuation of the modified, denoised transform is shown in Figure 2.14(c). By performing an inverse transform on the denoised transform, we produce the denoised whistle signal shown in Figure 2.14(d). When this denoised signal is played on a computer's sound system, the background static is completely gone and the volume of the pop noise, although still audible, is greatly diminished.

No claim is being made here that the method used for removing the pop noise is in any way optimal. There are two problems with the approach used. First, we only removed outliers from the fluctuation levels 2 and 4, and ignored higher levels. An examination of the transform reveals outliers for several other levels, such as 5 and 6. A more effective denoising would remove these spikes as well. Furthermore, actually removing the spikes completely may not be the best approach; reducing their size based on an estimate of nearby fluctuation values might work better. In any case, this example was described in order to indicate some of the flexibility available in wavelet denoising. More detailed discussions of denoising methods can be found in the references.

2.7 Two-dimensional wavelet transforms

Up till now we have been working with one-dimensional signals, but wavelet analysis can be done in any number of dimensions. The essential ideas, however, are revealed in two dimensions. In this section we shall begin our treatment of 2D wavelet analysis. Many of the basic ideas are similar to the 1D case; so we shall not repeat similar formulas but rather focus on the new ideas that are needed for the 2D case. In subsequent sections we shall describe various applications of 2D wavelet analysis to compression of images, denoising of images and other types of image enhancements and image analysis.

Discrete images

The 2D data that we shall be working with are discrete images. A *discrete image* $\mathbf{f}$ is an array of M rows and N columns of real numbers:

$$\mathbf{f} = \begin{pmatrix} f_{1,M} & f_{2,M} & \cdots & f_{N,M} \\ \vdots & \vdots & \ddots & \vdots \\ f_{1,2} & f_{2,2} & \cdots & f_{N,2} \\ f_{1,1} & f_{2,1} & \cdots & f_{N,1} \end{pmatrix}. \tag{2.39}$$

The *values* of $\mathbf{f}$ are the MN real numbers $\{f_{j,k}\}$. It should be noted that the way in which the values of $\mathbf{f}$ are displayed in the array on the right side of (2.39) is not the most commonly used one. We chose to display the values of $\mathbf{f}$ in this way because it corresponds well with the case where $\mathbf{f}$ is an array of sample values:

$$\mathbf{f} = \begin{pmatrix} g(x_1,y_M) & g(x_2,y_M) & \cdots & g(x_N,y_M) \\ \vdots & \vdots & \ddots & \vdots \\ g(x_1,y_2) & g(x_2,y_2) & \cdots & g(x_N,y_2) \\ g(x_1,y_1) & g(x_2,y_1) & \cdots & g(x_N,y_1) \end{pmatrix} \tag{2.40}$$

of a function $g(x, y)$ at the sample points (x_j, y_k) in the Cartesian coordinate plane. Just as with discrete 1D signals, it is frequently the case that a discrete 2D image is obtained from samples of some function $g(x, y)$.

It is often helpful to view a discrete image in one of two other ways. First, as a single column consisting of M signals having length N,

$$\mathbf{f} = \begin{pmatrix} \mathbf{f}_M \\ \vdots \\ \mathbf{f}_2 \\ \mathbf{f}_1 \end{pmatrix} \tag{2.41}$$

with the rows being the signals

$$\begin{aligned} \mathbf{f}_M &= (f_{1,M}, f_{2,M}, \ldots, f_{N,M}) \\ &\vdots \\ \mathbf{f}_2 &= (f_{1,2}, f_{2,2}, \ldots, f_{N,2}) \\ \mathbf{f}_1 &= (f_{1,1}, f_{2,1}, \ldots, f_{N,1}). \end{aligned}$$

Second, as a single row consisting of N signals of length M, written as columns,

$$\mathbf{f} = (\mathbf{f}^1, \mathbf{f}^2, \ldots, \mathbf{f}^N) \tag{2.42}$$

with the columns being the signals

$$\mathbf{f}^1 = \begin{pmatrix} f_{1,M} \\ \vdots \\ f_{1,2} \\ f_{1,1} \end{pmatrix}, \mathbf{f}^2 = \begin{pmatrix} f_{2,M} \\ \vdots \\ f_{2,2} \\ f_{2,1} \end{pmatrix}, \ldots, \mathbf{f}^N = \begin{pmatrix} f_{N,M} \\ \vdots \\ f_{N,2} \\ f_{N,1} \end{pmatrix}.$$

Notice that, because of our somewhat peculiar notation, the row index for each column increases from bottom to top (rather than from top to bottom, which is more common notation in image processing).

As an example of the utility of Formulas (2.41) and (2.42), we consider the calculation of the energy of a discrete image. The *energy* $\mathcal{E}_{\mathbf{f}}$ of a discrete image $\mathbf{f}$ is defined to be the sum of the squares of all of its values. Because of (2.41) it follows that $\mathcal{E}_{\mathbf{f}}$ is the sum of the energies of all of the row signals:

$$\mathcal{E}_{\mathbf{f}} = \mathcal{E}_{\mathbf{f}_1} + \mathcal{E}_{\mathbf{f}_2} + \cdots + \mathcal{E}_{\mathbf{f}_M}.$$

Or, because of (2.42), it follows that $\mathcal{E}_{\mathbf{f}}$ is also the sum of the energies of all of the column signals:

$$\mathcal{E}_{\mathbf{f}} = \mathcal{E}_{\mathbf{f}^1} + \mathcal{E}_{\mathbf{f}^2} + \cdots + \mathcal{E}_{\mathbf{f}^N}.$$

One consequence of these last two identities is that the 2D wavelet transforms defined below have the property of conserving the energy of discrete images.

2D wavelet transforms

A 2D wavelet transform of a discrete image can be performed whenever the image has an even number of rows and an even number of columns. A 1-level wavelet transform of an image $\mathbf{f}$ is defined, using any of the 1D wavelet transforms that we have discussed, by performing the following two steps:

Step 1. Perform a 1-level, 1D wavelet transform, on each row of $\mathbf{f}$, thereby producing a new image.

Step 2. On the new image obtained from Step 1, perform the same 1D wavelet transform on each of its columns.

It is not difficult to show that Steps 1 and 2 could be done in reverse order and the result would be the same. A 1-level wavelet transform of an image $\mathbf{f}$ can be symbolized as follows:

$$\mathbf{f} \longmapsto \begin{pmatrix} \mathbf{h}^1 & | & \mathbf{d}^1 \\ — & & — \\ \mathbf{a}^1 & | & \mathbf{v}^1 \end{pmatrix} \tag{2.43}$$

where the subimages $\mathbf{h}^1$, $\mathbf{d}^1$, $\mathbf{a}^1$, and $\mathbf{v}^1$ each have $M/2$ rows and $N/2$ columns. We shall now discuss the nature of each of these subimages.

The subimage $\mathbf{a}^1$ is created by computing trends along rows of $\mathbf{f}$ followed by computing trends along columns; so it is an averaged, lower resolution version of the image $\mathbf{f}$. For example, in Figure 2.15(a) we show a simple test image of an octagon, and in Figure 2.15(b) we show its 1-level Coif6 transform. The $\mathbf{a}^1$ subimage appears in the lower left quadrant of the Coif6 transform, and it is clearly a lower resolution version of the original octagon image. Since a 1D trend computation is $\sqrt{2}$ times an average of successive values in a signal, and the 2D trend subimage $\mathbf{a}^1$ was computed from trends along both rows and columns, it follows that each value of $\mathbf{a}^1$ is equal to 2 times an average of a small square containing adjacent values from the image $\mathbf{f}$. A useful way of expressing the values of $\mathbf{a}^1$ is as scalar products of the image $\mathbf{f}$ with scaling signals, as we did in the 1D case; we shall say more about this later in the section.

The $\mathbf{h}^1$ subimage is created by computing trends along rows of the image $\mathbf{f}$ followed by computing fluctuations along columns. Consequently, wherever there are horizontal edges in an image, the fluctuations along columns are able to detect these edges. This tends to emphasize the horizontal edges, as can be seen clearly in Figure 2.15(b) where the subimage $\mathbf{h}^1$ appears in the upper left quadrant. Furthermore, notice that vertical edges, where the octagon image is constant over long stretches, are removed from the subimage $\mathbf{h}^1$. This discussion should make it clear why we shall refer to this subimage as the first *horizontal fluctuation.*

The subimage $\mathbf{v}^1$ is similar to $\mathbf{h}^1$, except that the roles of horizontal and vertical are reversed. In Figure 2.15(b) the subimage $\mathbf{v}^1$ is shown in the lower right quadrant. Notice that horizontal edges of the octagon are erased, while vertical edges are emphasized. This is typically the case with $\mathbf{v}^1$, which we shall refer to as the first *vertical fluctuation.*

Finally, there is the first *diagonal fluctuation,* $\mathbf{d}^1$. This subimage tends to emphasize diagonal features, because it is created from fluctuations along both rows and columns. These fluctuations tend to erase horizontal and vertical edges where the image is relatively constant. For example, in Figure 2.15(b) the diagonal fluctuation appears in the upper right quadrant of the image, and it is clear that diagonal details are emphasized while horizontal and vertical edges are erased.

It should be noted that the basic principles discussed previously for 1D wavelet analysis still apply here in the 2D setting. For example, the fact that fluctuation values are generally much smaller than trend values is still true. In the wavelet transform shown in Figure 2.15(b), for instance, the fluctuation subimages $\mathbf{h}^1$, $\mathbf{v}^1$, and $\mathbf{d}^1$ have significantly smaller values than the values in the trend subimage $\mathbf{a}^1$. In fact, in order to make the values for $\mathbf{h}^1$, $\mathbf{v}^1$, and $\mathbf{d}^1$ visible, they are displayed on a logarithmic intensity scale, while the values for the trend subimage $\mathbf{a}^1$ are displayed using an ordinary, linear scale.

Furthermore, 2D wavelet transforms enjoy the Conservation of Energy property. As noted above, the energy of an image is the sum of the energies of each of its rows or each of its columns. Since the 1D wavelet transforms of the rows, performed in Step 1, preserve the row energies, the image obtained in Step 1 will have the same energy as the original image. Likewise, since the 1D wavelet transforms of the columns preserve their energies, it follows that the transform obtained in Step 2 has the same energy as the image from Step 1. Thus the 1-level wavelet transform has the same energy as the original image. For example, the energy of the octagon image in Figure 2.15(a) is 3919.0625, while the energy of its 1-level Coif6 transform is 3919.0622; the slight discrepancy between the two energies is attributable to the inevitable rounding error that arises from finite precision computer calculations.

As in 1D, multiple levels of 2D wavelet transforms are defined by repeating the 1-level transform of the previous trend. For example, a 2-level wavelet transform is performed by computing a 1-level transform of the trend subimage $\mathbf{a}^1$ as follows:

$$\mathbf{a}^1 \longmapsto \begin{pmatrix} \mathbf{h}^2 & | & \mathbf{d}^2 \\ - & & - \\ \mathbf{a}^2 & | & \mathbf{v}^2 \end{pmatrix}.$$

The 1-level fluctuations $\mathbf{h}^1$, $\mathbf{d}^1$, and $\mathbf{v}^1$ remain unchanged. In Figure 2.15(c) we show a 2-level Coif6 transform of the octagon image. In general, a k-level transform is defined by performing a 1-level transform on the previous trend $\mathbf{a}^{k-1}$. In Figure 2.15(d) we show a 3-level Coif6 transform of the octagon image.

It is interesting to compare the successive levels of the Coif6 transforms in Figure 2.15. Notice how it appears that we are systematically decomposing the original octagon image by peeling off edges; and these edges are retained within the fluctuation subimages. This aspect of wavelet transforms plays a major role in the fields of *image recognition* and *image enhancement.* In Section 2.11 we shall discuss a few examples from these fields.

Besides Conservation of Energy these 2D wavelet transforms also perform a Compaction of Energy. For example, for the octagon image in Figure 2.15 most of the energy of the image is successively localized into smaller and smaller trend subimages, as summarized in Table 2.7. Notice, for example, that the third trend $\mathbf{a}^3$, which is 64 times smaller than $\mathbf{f}$ in terms of numbers of values, still contains over 96% of the total energy. In accordance with the Uncertainty Principle, however, some of the energy has leaked out into the fluctuation subimages. Consequently, in order to obtain an accurate approximation of $\mathbf{f}$, some of the highest energy fluctuation values—such as the ones that are visible in Figure 2.15(d)—would have to be included along with the third trend values when performing an inverse transform. This Compaction of Energy property, as in the 1D case, provides the foundation for the methods of compression and denoising that we shall discuss in subsequent sections.

Table 2.7 Compaction of energy of octagon image

Image	*Energy*	*% Total Energy*
$\mathbf{f}$	3919.06	100.00
$\mathbf{a}^1$	3830.69	97.75
$\mathbf{a}^2$	3811.06	97.22
$\mathbf{a}^3$	3777.55	96.39

2D wavelets and scaling images

As in the 1D case, the various levels of a wavelet transform can be computed via scalar products of the image $\mathbf{f}$ with elementary images called *scaling images* and *wavelets.* A scalar product of two images $\mathbf{f}$ and $\mathbf{g}$, both having M rows and N columns, is defined by

$$\mathbf{f} \cdot \mathbf{g} = f_{1,1}g_{1,1} + f_{1,2}g_{1,2} + \cdots + f_{N,M}g_{N,M}. \tag{2.44}$$

In other words, $\mathbf{f} \cdot \mathbf{g}$ is the sum of all the products of similarly indexed values of $\mathbf{f}$ and $\mathbf{g}$.

To see how this scalar product operation relates to wavelet transforms, let's consider the 1-level horizontal fluctuation $\mathbf{h}^1$. This subimage is defined by separate calculations of trends along rows and fluctuations along columns. It follows that the values of $\mathbf{h}^1$ are computed via scalar products with wavelets obtained by multiplying values of 1D scaling signals along

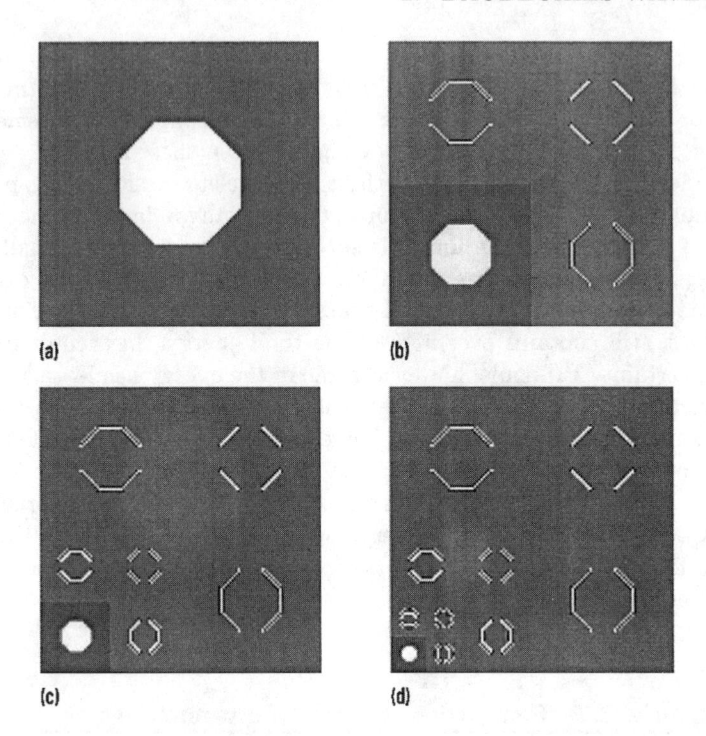

FIGURE 2.15
(a) Octagon image. (b) 1-level Coif6 transform. (c) 2-level Coif6 transform. (d) 3-level Coif6 transform.

rows by values of 1D wavelets along columns. Each such wavelet is denoted by $\mathbf{V}_m^1 \otimes \mathbf{W}_n^1$, which is called a *tensor product* of the 1D scaling signal $\mathbf{V}_m^1$ and 1D wavelet $\mathbf{W}_n^1$. For instance, if we are computing a 2D Haar wavelet transform, then $\mathbf{V}_1^1 \otimes \mathbf{W}_1^1$ is defined as follows:

$$\mathbf{V}_1^1 \otimes \mathbf{W}_1^1 = \begin{pmatrix} 0 & 0 & 0 & \dots & 0 & 0 \\ 0 & 0 & 0 & \dots & 0 & 0 \\ \vdots & \vdots & \vdots & \ddots & \vdots & \vdots \\ 0 & 0 & 0 & \dots & 0 & 0 \\ -1/2 & -1/2 & 0 & \dots & 0 & 0 \\ 1/2 & 1/2 & 0 & \dots & 0 & 0 \end{pmatrix}.$$

Since each Haar scaling signal $\mathbf{V}_m^1$ is a translation by $2(m-1)$ time-units of $\mathbf{V}_1^1$, and each Haar wavelet $\mathbf{W}_n^1$ is a translation by $2(n-1)$ time-units of $\mathbf{W}_1^1$, it follows that $\mathbf{V}_m^1 \otimes \mathbf{W}_n^1$ is a translation by $2(m-1)$ units along the horizontal and $2(n-1)$ units along the vertical of $\mathbf{V}_1^1 \otimes \mathbf{W}_1^1$. Notice that the Haar wavelet $\mathbf{V}_1^1 \otimes \mathbf{W}_1^1$ has energy 1 and an average value of 0, as do all the other Haar wavelets $\mathbf{V}_m^1 \otimes \mathbf{W}_n^1$. Furthermore, the support of the Haar wavelet $\mathbf{V}_1^1 \otimes \mathbf{W}_1^1$ is a 2 by 2 square, and so the support of each

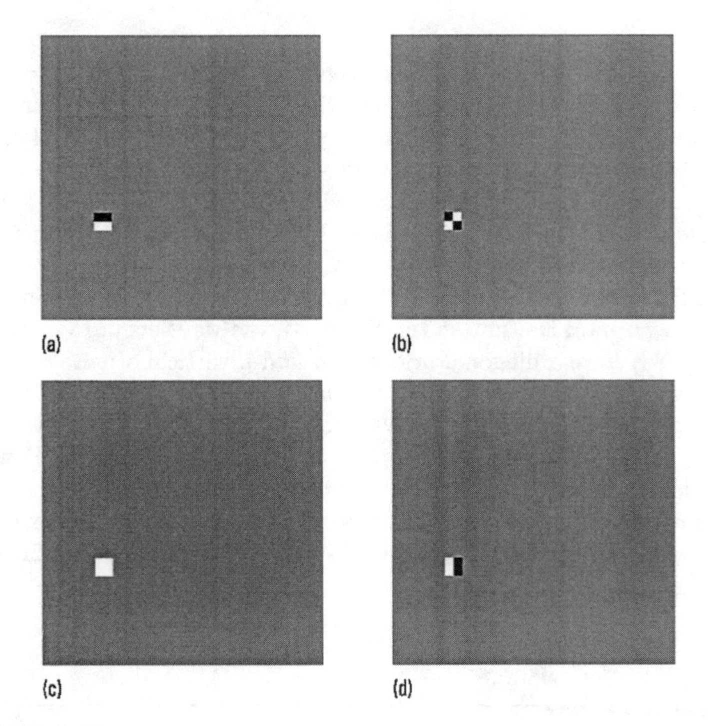

FIGURE 2.16
(a) Haar wavelet $\mathbf{V}_3^2 \otimes \mathbf{W}_5^2$. (b) Haar wavelet $\mathbf{W}_3^2 \otimes \mathbf{W}_5^2$. (c) Haar scaling image $\mathbf{V}_3^2 \otimes \mathbf{V}_5^2$. (d) Haar wavelet $\mathbf{W}_3^2 \otimes \mathbf{V}_5^2$. Note: the gray background indicates zero values, white indicates positive values, and black indicates negative values.

Haar wavelet $\mathbf{V}_m^1 \otimes \mathbf{W}_n^1$ is a 2 by 2 square as well. For the Daubechies wavelets, the supports of the wavelets $\mathbf{V}_m^1 \otimes \mathbf{W}_n^1$ are also small squares, although not 2 by 2 ones. The Daub4 wavelets $\mathbf{V}_m^1 \otimes \mathbf{W}_n^1$, for instance, all have supports in some 4 by 4 square (if we allow for wrap-around at image boundaries).

Similar definitions hold for the other subimages of the 1-level transform. The values of the diagonal fluctuation $\mathbf{d}^1$ are scalar products of the image with the wavelets $\mathbf{W}_m^1 \otimes \mathbf{W}_n^1$, and the values of the vertical fluctuation $\mathbf{v}^1$ are scalar products of the image with the wavelets $\mathbf{W}_m^1 \otimes \mathbf{V}_n^1$. All of these wavelets have energy 1 and average value 0, and supports in small squares. The values of the trend $\mathbf{a}^1$ are scalar products of the image with the scaling images $\mathbf{V}_m^1 \otimes \mathbf{V}_n^1$. Each of these scaling images has energy 1 and average value of 1/2, and support in some small square.

What is true for the first level remains true for all subsequent levels. The values of each subimage $\mathbf{a}^k$, $\mathbf{h}^k$, $\mathbf{d}^k$, and $\mathbf{v}^k$ are computed by scalar products with the scaling images $\mathbf{V}_m^k \otimes \mathbf{V}_n^k$, and the wavelets $\mathbf{V}_m^k \otimes \mathbf{W}_n^k$,

$\mathbf{W}_m^k \otimes \mathbf{W}_n^k$, and $\mathbf{W}_m^k \otimes \mathbf{V}_n^k$, respectively. In Figure 2.16 we show graphs of a 2-level Haar scaling image and three 2-level Haar wavelets. Notice how these images correspond with the subimages of the 2-level Haar transform. For instance, the scaling image $\mathbf{V}_3^2 \otimes \mathbf{V}_5^2$, shown in Figure 2.16(c), is supported on a small 4 by 4 square with a constant value of 1/4, and the average of an image over this square produces the trend value in the $(3, 5)$ position in the trend subimage $\mathbf{a}^2$. Also notice the horizontal orientation of the Haar wavelet in Figure 2.16(a), which is used for computing a value for the horizontal fluctuation $\mathbf{h}^2$. Similarly, the wavelets in Figures 2.16(b) and 2.16(d) have a diagonal orientation and a vertical orientation, respectively, which is consistent with their being used for computing values for the diagonal fluctuation $\mathbf{d}^2$ and vertical fluctuation $\mathbf{v}^2$.

This concludes a very brief outline of the basic features of 2D wavelets and scaling images. Their essential features are very similar to the 1D case; so for reasons of space we shall limit ourselves to this brief treatment. In the next section we shall begin our discussion of applications of 2D wavelet analysis.

2.8 Compression of images

In this section we begin our treatment of image processing applications with a brief treatment of wavelet compression of images. The basic ideas involved are similar to the methods described previously, in Sections 2.4 and 2.5, for compression of 1D signals. We shall concentrate on one representative example of image compression, which should indicate many of the exciting features of wavelet compression techniques, as well as some of the significant challenges in this new field. In the section that follows we shall expand on some of the ideas introduced here by describing a fascinating subfield of image compression: the compression of fingerprint images.

The image that we shall compress is shown in Figure 2.17(a). It is a standard test image, known as *Lena,* which appears frequently in the field of image processing. While one reason for this may be the attractiveness of the image to many of the male workers in the field, another more serious reason is that the *Lena* image contains various combinations of image properties, such as a large number of curved edges and textures in *Lena's* hair and the feathers on her hat, and combinations of light and dark regions of shading in the background and in *Lena's* face.

The *Lena* image is a 512 row by 512 column discrete image. Its values are gray-scale intensity values from 0 to 255—that is, quantized values at 8 bpp—where 0 indicates pure black and 255 indicates pure white, and other values indicate shades of gray between these two extremes. To compress

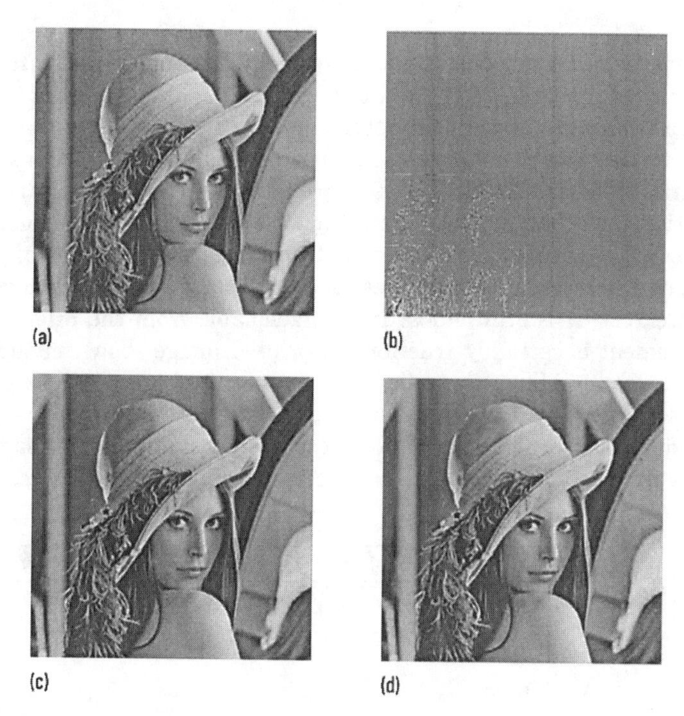

FIGURE 2.17
(a) *Lena* image. (b) 4-level Coif12 transform. (c) Compressed image, 0.43 bpp. (d) Compressed image, 0.24 bpp.

the *Lena* image we shall use its 4-level Coif12 transform; this transform is shown in Figure 2.17(b). It is clear from this figure that the first level fluctuations—$\mathbf{h}^1$, $\mathbf{d}^1$, and $\mathbf{v}^1$—contain very little energy. An elementary compression could be done by simply omitting all these fluctuations from the compressed version. This is equivalent to compressing the *Lena* image $\mathbf{f}$ to its first level trend $\mathbf{a}^1$, hence compressing *Lena* to 2 bpp.

Much more compression can be obtained, however, if we modify the simple compression described above along the lines of the compression method described for the 1D signal, *greasy,* at the end of Section 2.5. In Figure 2.17(c) we show a compression obtained by quantizing the fourth trend subimage, $\mathbf{a}^4$, of the 4-level Coif12 transform at 8 bpp, and quantizing the second, third, and fourth fluctuations at 7 bpp, and omitting the first fluctuations entirely. Computing separate entropies for each level, we obtain an estimate of 0.43 bpp needed to encode the significant values in the transform. Since the significance map in this case consists of bits of 1's and 0's for only the lower left quadrant of the quantized transform, it accounts for 0.25 bpp without compression of any kind. Examining Figure 2.17(b), however, reveals large areas that consist only of 0's. Therefore, the significance map can be significantly compressed. The image in Figure 2.17(c) is virtually indistinguishable from the original *Lena* image, even though it has been compressed by roughly 16:1.

Even further compression of the *Lena* image is possible. For instance, if the fourth trend is quantized at 9 bpp and the second, third, and fourth fluctuations are quantized at 6 bpp, then the estimated average number of bits needed to encode the significant transform values is 0.24 bpp. This represents a much greater compression; yet, the compressed image shown in Figure 2.17(d) is still almost indistinguishable from the original. This last statement is certainly true for the printed image shown in the figure; however, when the images are displayed on a computer screen, the 0.24 bpp compressed image exhibits some defects in comparison to the original.

One quantitative measure of how accurately a compressed image $\mathbf{g}$ approximates the original image $\mathbf{f}$ is the *relative 2-norm difference* $\mathcal{D}(\mathbf{f},\mathbf{g})$ defined by

$$\mathcal{D}(\mathbf{f},\mathbf{g}) = \frac{\sqrt{(f_{1,1}-g_{1,1})^2+(f_{1,2}-g_{1,2})^2+\cdots+(f_{N,M}-g_{N,M})^2}}{\sqrt{f_{1,1}^2+f_{1,2}^2+\cdots+f_{N,M}^2}}$$

$$= \sqrt{\mathcal{E}_{\mathbf{f}-\mathbf{g}}/\mathcal{E}_{\mathbf{f}}}\,. \tag{2.45}$$

For example, if $\mathbf{f}$ is the *Lena* image and $\mathbf{g}$ is the 0.43 bpp compressed image in Figure 2.17(c), then $\mathcal{D}(\mathbf{f},\mathbf{g}) = 0.040$. While if $\mathbf{g}$ is the 0.25 bpp compressed image in Figure 2.17(d), then $\mathcal{D}(\mathbf{f},\mathbf{g}) = 0.046$. This gives quantitative support to the subjective impression that the compressed image in Figure 2.17(c) is a closer approximation to the original image.

As a rule of thumb, *if* $\mathcal{D}(\mathbf{f},\mathbf{g}) \leq .05$, *then* $\mathbf{g}$ *is an acceptable approximation to* $\mathbf{f}$. This is certainly true for the images in Figure 2.17. The use of $\mathcal{D}(\mathbf{f},\mathbf{g})$ as a measure of acceptable approximation, however, does not always equate with the perceptions of our visual systems. No such quantitative measure is known, although interesting proposals based on the magnitudes of wavelet transform values have been made in recent papers that we list in the references.

These compressions of the *Lena* image are hardly state of the art. By combining the best available techniques of entropy encoding of the significant quantized transform values with a sophisticated method of compressing the significance map known as *zero-tree encoding,* it is possible to compress *Lena* at a ratio of at least 50:1 without noticeable degradation. The foundation of the zero-tree method is relatively simple. It rests on the fact that trend images are accurate reproductions, at lower resolutions, of the original image. Consequently, in the *Lena* image for instance, regions such as her shoulder which have relatively constant intensity[7] will produce in-

[7]Or for which intensity varies along a flat, linear gradient.

significant values *in the same relative locations* at several consecutive levels. Hence, an insignificant transform value in, say $\mathbf{h}^3$, will often correspond to *four* insignificant values in $\mathbf{h}^2$ in the same relative location (relative to the original image). Likewise, each of those insignificant values in $\mathbf{h}^2$ will often correspond to four more insignificant values, 16 in all, in the same relative locations in $\mathbf{h}^1$. Consequently, these values can be grouped in a data structure, called a *zero-tree.* Encoding these zero-trees with just a single bit is analogous to run-length encoding of zeros, but is much better correlated to the structure of the significance map and produces phenomenal compression ratios. There are very lucid discussions of these matters in some of the original papers which we list in the references. We shall also give some further discussion of zero-trees at the end of the next section.

One aspect of image compression that our brief introduction has omitted is the compression of color images. The way in which most color images are encoded, as Red-Green-Blue (RGB) intensities, makes it possible to consider their compression as a relatively straightforward generalization of compression of gray-scale images. The reason for this is that the total *intensity,* I, which equals the average of the R, G, and B intensities, has much more effect on the visual perception of the color image than the two color values of *hue,* H, and *saturation,* S. Consequently, the RGB image is mapped to an IHS image before compression is done. (The formulas for this mapping are described in the references for this chapter.) After performing this mapping the I, H, and S values are compressed as separate images, using the methods for gray-scale compression illustrated above. Much greater compression can be done on the H and S images than on the I image because of the much lower sensitivity that the human visual system has for variations in hue and saturation. More details on color compression can be found in the references for this chapter.

2.9 Fingerprint compression

Fingerprint compression is an interesting case of image compression. In this section we shall briefly outline the essential ideas underlying wavelet compression of fingerprints. A wavelet-based compression algorithm has been adopted by the U.S. government, in particular by the FBI, as its standard for transmitting and storing digitized fingerprints.

A *fingerprint image* is an image of a fingerprint in an 8-bit gray-scale format. A typical fingerprint record—consisting of ten fingerprints plus two additional thumbprints and two full handprints—when digitized as images produces about 10 megabytes of data. This magnitude of data poses significant problems for transmission and storage. For example, to transmit

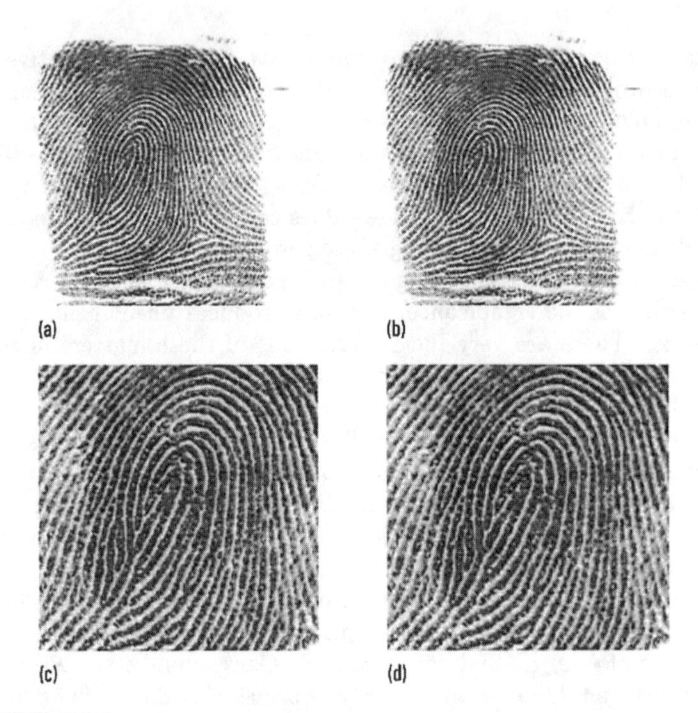

FIGURE 2.18
(a) Fingerprint 1, 8 bpp. (b) Compressed version, 0.77 bpp. (c) Central whorl in (a). (d) Central whorl in (b).

one fingerprint record over a modem, say at 28,000 bits per second with an overhead of around 20%, would require about 1 hour. This is painfully slow when identification needs to be done quickly, as is often the case in criminal investigations. If fingerprint images could be compressed at, say, a 20:1 ratio *without noticeable loss of detail,* then the transmission of one fingerprint record could be reduced to just 3 minutes. This would greatly facilitate the large number of background checks (around 30,000 each day) that are needed.

Besides the transmission problem, there are also problems stemming from the gargantuan magnitude of the FBI's fingerprint archive. This archive contains over 25 million records. Digitizing these records at 8 bpp would produce over 250 trillion bytes of data! Compressing each image by a factor of 20:1 would greatly ease the storage burden for an archive of these fingerprint images.

The wavelet-based compression algorithm adopted by the U.S. government is called the *Wavelet/Scalar Quantization* (WSQ) method. We will not try to give a complete description of the WSQ method. Rather, we shall describe enough of its essential aspects, so that the reader should be able to confidently read some of the excellent articles by the creators of the WSQ method. These articles are listed in the Notes and References section at the end of this chapter.

In this section, we illustrate the gist of the WSQ method by discussing an example of a fingerprint compression using a wavelet transform. This initial example will be expanded in Chapter 4, in order to give further insight into the WSQ method.

Consider the test image, *Fingerprint 1,* shown in Figure 2.18(a). To compress this image we used a 4-level Coif18 transform. The fourth trend $\mathbf{a}^4$ was quantized at 9 bpp and the various fluctuations were quantized at 6 bpp. This produces an estimated 0.77 bpp needed to encode the significant transform values. Compared with 8 bpp needed for the original image, this represents roughly 10:1 compression. Here we are ignoring the bits needed to encode the significance map; we shall give some justification for doing so at the end of this section. The compressed image is shown in Figure 2.18(b). This image is virtually indistinguishable from the original image in (a).

As a quantitative measure of the accuracy of the compressed image, we calculate the relative 2-norm difference $\mathcal{D}(\mathbf{f}, \mathbf{g})$ defined in the previous section. If $\mathbf{f}$ is the Fingerprint 1 image and $\mathbf{g}$ is the compressed image, then $\mathcal{D}(\mathbf{f}, \mathbf{g}) \doteq .035$. This value is significantly less than the accuracy value of .05 proposed in the previous section.

Often portions of fingerprints need to be magnified in order to compare certain details, such as *whorls.* In Figures 2.18(c) and (d) we show magnifications of the central whorl from Fingerprint 1 and from its compressed version. The compressed version is still virtually indistinguishable from the original in these magnifications. For these magnifications the relative 2-norm difference is .054, which is very near to the accuracy value of .05. It is interesting to note that even the tiny indentations in the middle of the fingerprint ridges, which correspond to sweat pores, are accurately preserved by the compressed image. The locations of these sweat pores are, in fact, legally admissible for identification purposes.

While this example illustrates that a wavelet transform compression of fingerprints can work reasonably well, we can do even better. In Chapter 4, we will show that a *wavelet packet* transform of Fingerprint 1 produces a significantly greater compression. The WSQ method is essentially just a slight modification of this wavelet packet method.

Remarks on the significance map

In our discussion above of a wavelet compression of Fingerprint 1, we ignored the presence of the significance map. In this subsection we will explain why the significance map can be greatly compressed, and consequently does not contribute much to the number of bits needed to describe the compressed fingerprint image.

We begin by noting that only 12.7% or about one eighth of all of the transform values for Fingerprint 1 are significant. Hence there will be a large preponderance of zero bits in the significance map, which implies that it should be very compressible.

More important, however, many of these zero bits are arranged in the zero-tree data structures mentioned at the end of the previous section. To see this, we show the significance map for the compression of Fingerprint 1 in Figure 2.19. A careful inspection of this significance map reveals many such zero-trees just by sight alone. We start by locating the significance bits for the fourth trend $\mathbf{a}^4$ as the small white square in the lower left corner of Figure 2.19. Directly above the top right corner of this square is a sequence of several gray blobs; these represent clusters of zero bits at the right edge of the fourth horizontal fluctuation $\mathbf{h}^4$. Corresponding to each one of these zero bits are four zero bits, in the same relative location in the third horizontal fluctuation $\mathbf{h}^3$. Thinking of this as an expansion of the area of the blobs by a factor of four, we can see that there is room for these larger blobs lying within a gray rectangle at the edge of the third horizontal fluctuation. Moreover, this entire gray rectangle expands by a factor of four to produce another gray rectangle lying along the side of the second horizontal fluctuation, and this second gray rectangle expands by a factor of four to produce a third gray rectangle along the side of the first horizontal fluctuation.

We have thus found that each zero bit in the original set of gray blobs lies at the root of a four-level zero-tree composed of 85 zero bits. Each of these zero-trees can be encoded with one symbol, say $\mathcal{Z}$, which is in effect an 85:1 compression of each of these zero-trees. Furthermore, each of the extra zero bits within the gray rectangle at the side of the third horizontal fluctuation (the ones not contained within the four-level zero-trees) lies at the root of a three-level zero-tree composed of 20 zero bits. Each of these zero-trees can be collapsed to a symbol $\mathcal{Z}$, producing in effect a 20:1 compression of these trees.

The reader might find it amusing to visually locate other zero-trees. They make up a substantial percentage of all the zero bits in the significance map, and allow for high compressibility.

Finally, we note that the method of zero-tree compression encodes the signs of the significant transform values as symbols (say $\mathcal{P}$ and $\mathcal{N}$) within the significance map. That is, instead of 1 for a significant transform value, a symbol of $\mathcal{P}$ for positive or $\mathcal{N}$ for negative is used instead. Essentially no extra bits are needed for this, because the sign bits can be dropped from the encoding of the significant transform values, leaving only their absolute values to be encoded. Often this significantly decreases the number of bits needed for encoding the significant transform values. For example, for Fingerprint 1, if only the absolute values of the significant 4-level Coif18

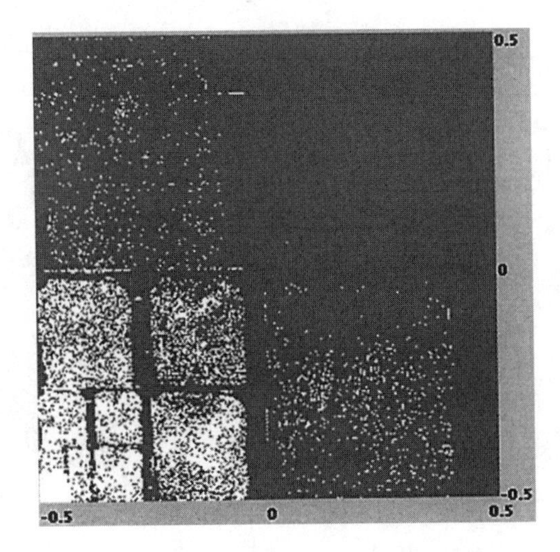

FIGURE 2.19
Significance map for compressed fingerprint. Gray pixels are zero bits, white zeros are one bits.

transform values are encoded, then only 0.64 bpp are needed. This is a 17% reduction from the 0.77 bpp needed when sign bits were included.

For the reasons described above, the significance map often compresses to such a large degree that (taking into account the decrease of bits needed when sign bits are ignored) we can safely ignore it when making a rough estimate of the bpp needed in a wavelet compression of fingerprints.

2.10 Denoising images

In this section we shall describe some fundamental wavelet based techniques for removing noise from images.[8] Noise removal is an essential element of image processing. One reason for this is that many images are acquired under less than ideal conditions and consequently are contaminated by significant amounts of noise. This is the case, for example, with many medical images. Another reason is that several important image processing operations, such as contrast enhancement, histogram equalization,

[8]Adopting audio terminology, undesired changes in image values are called *noise*.

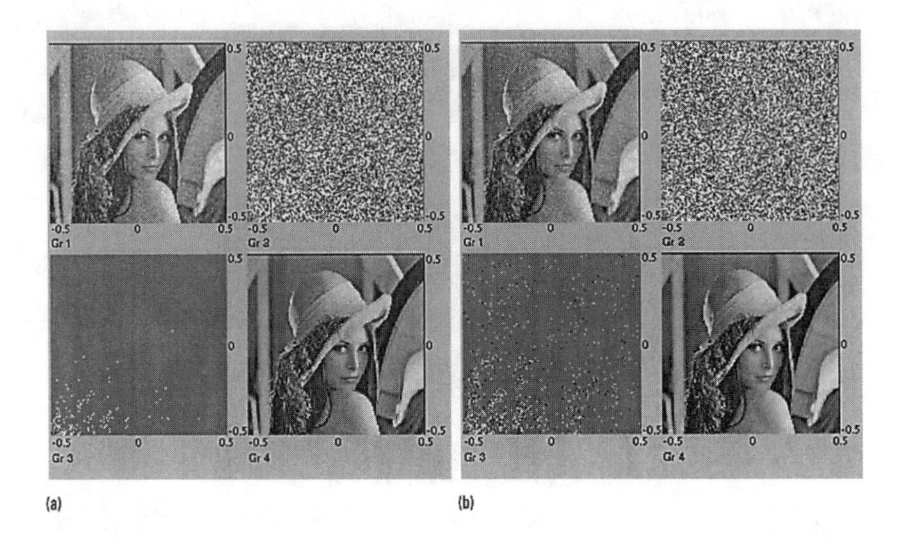

FIGURE 2.20
(a) Hard threshold denoising: Gr 1, *noisy Lena;* Gr 2, 4-level Coif12 transform; Gr 3, hard thresholded transform; Gr 4, denoised image. (b) Soft threshold denoising; the soft thresholded transform is Gr 3.

and edge enhancement, work much better if random noise is absent.

The basic concepts of wavelet denoising of images are similar to those described previously for 1D audio signals in Section 2.6. By choosing a threshold that is a sufficiently large multiple of the standard deviation of the random noise, it is possible to remove most of the noise by thresholding of wavelet transform values. We shall examine how this procedure, which is known as *hard thresholding,* works on a noisy version of the *Lena* image. A simple modification of this method, known as *soft thresholding,* will also be introduced and compared with hard thresholding. We shall then conclude the section with a brief discussion of further aspects of wavelet denoising, which should indicate some of the variety of techniques that wavelet analysis provides for denoising images.

Our first example of image denoising is a hard threshold denoising of a noisy version of the *Lena* image. This image, which we shall refer to as *noisy Lena,* is shown at the top left of Figure 2.20(a). It was obtained by adding Gaussian random noise to the *Lena* image, shown in Figure 2.17(a).

We saw in the previous section that the *Lena* image can be effectively compressed using a 4-level Coif12 transform. This was because there were relatively few, high-energy transform values which captured most of the energy of the image. Therefore, a 4-level Coif12 transform should be an effective tool for denoising the *noisy Lena* image. At the upper right of

Figure 2.20(a) we show a 4-level Coif12 transform of the *noisy Lena* image. An estimate was made of the standard deviation σ of the noise values, using values in the central portion of the diagonal fluctuation $\mathbf{d}^1$. The threshold value T was then set at 4.5σ as per Formula (2.38). At the lower left of Figure 2.20(a) we show the hard thresholded transform; all transform values whose magnitudes are less than T are set equal to 0, while the remaining transform values are retained as significant values. An inverse transform was performed on this thresholded transform, producing the denoised image at the lower right in Figure 2.20(a).

The denoised *Lena* image is clearly an improvement over the noisy version. The RMS Error between the denoised image and the uncontaminated *Lena* image is 15.8. This shows a small reduction from the RMS Error of 17.8 for the *noisy Lena* image, but the amount of reduction does not seem to be large enough to accurately reflect the perceived reduction of noise.[9] Later in this section, we shall discuss an alternative measure which better reflects the amount of noise reduction.

We shall now describe the soft thresholding method of denoising. Soft thresholding is a simple modification of hard thresholding. Hard thresholding consists of applying the following function

$$H(x) = \begin{cases} x & \text{if } |x| \geq T \\ 0 & \text{if } |x| < T \end{cases} \tag{2.46}$$

to the wavelet transform values. See Figure 2.21(a). As can be seen in this figure, the hard threshold function H is not continuous, and thus greatly exaggerates small differences in transform values whose magnitudes are near the threshold value T. If a value's magnitude is only slightly less than T, then this value is set equal to 0, while a value whose magnitude is only slightly greater than T is left unchanged. Soft thresholding replaces the discontinuous function H by a continuous function S, such as[10]

$$S(x) = \begin{cases} x & \text{if } |x| \geq T \\ 2x - T & \text{if } T/2 \leq x < T \\ T + 2x & \text{if } -T < x \leq -T/2 \\ 0 & \text{if } |x| < T/2. \end{cases} \tag{2.47}$$

See Figure 2.21(b). This soft threshold function S does not exaggerate the gap between significant and insignificant transform values.

In Figure 2.20(b) we show a soft threshold denoising of the *noisy Lena* image, using a 4-level Coif12 transform, and also using the same threshold value T as for the hard thresholding in Figure 2.21(a). The soft thresholded transform is shown at the lower left of Figure 2.18(b). It can be

[9]This may not be apparent from the printed images, in which case we urge the reader to examine these images on a computer display using FAWAV.

[10]The function in (2.47) is not the only soft threshold function that is used.

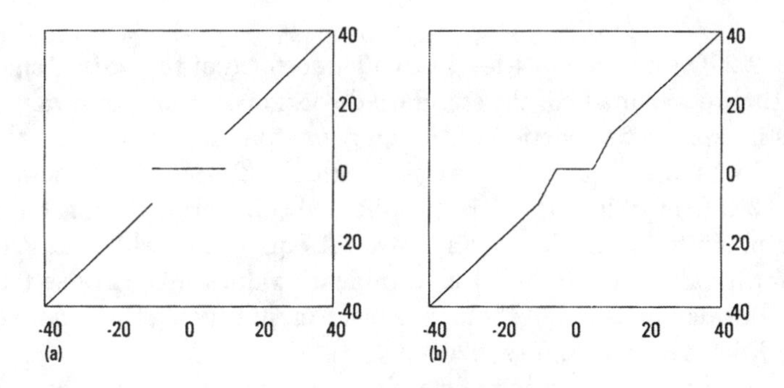

FIGURE 2.21
(a) Hard threshold function, $T = 10$. (b) Soft threshold function, $T = 10$.

clearly seen that the soft thresholding retains more transform values than the hard thresholding. Applying the inverse transform to the soft thresholded transform produces the denoised image at the lower right of Figure 2.21(b). Although it is difficult to distinguish the printed versions of the two denoised images, the computer displayed versions—which are accessible from the figure files available from the FAWAV website—are clearly distinguishable, with the soft threshold denoising appearing to be slightly superior. The RMS Error confirms this subjective judgment. For the soft threshold denoising the RMS Error is 15.20, which is slightly less than the RMS Error of 15.80 for the hard threshold denoising.

Quantitative measures of error

There are several ways of measuring the amount of error between a noisy image and the original image. All of these measures aim to provide quantitative evidence for the effectiveness of noise removal.

One measure is RMS Error, which we used above. Unfortunately, it did not seem to accurately reflect the perceived substantial decrease in noise in either the hard threshold or soft threshold denoising of the *noisy Lena* image.

Another measure is the relative 2-norm difference, $\mathcal{D}(\mathbf{f}, \mathbf{g})$, defined in Formula (2.45). Here $\mathbf{f}$ is the original, uncontaminated image and $\mathbf{g}$ is a noisy image. The errors calculated using $\mathcal{D}$ with the *noisy Lena* example are summarized in Table 2.8. Again, as with RMS Error, this measure of error $\mathcal{D}$ does not seem to accurately reflect the amount of perceived denoising. These two measures, RMS Error and relative 2-norm difference, provide essentially the same information. In fact, simple algebra shows that the ratios of different errors—which are computed in order to compare percentages of denoising—are always the same for these two measures.

A third measure, commonly used in image processing, is *Signal to Noise Ratio* (SNR). If $\mathbf{f}$ is the original image and $\mathbf{g}$ is a noisy image, then the SNR measure of error is defined by

$$\begin{aligned} SNR &= 20 \log_{10} [1/\mathcal{D}(\mathbf{f}, \mathbf{g})] \\ &= 10 \log_{10} [\mathcal{E}_{\mathbf{f}}/\mathcal{E}_{\mathbf{g}-\mathbf{f}}] . \end{aligned} \tag{2.48}$$

A rationale for using SNR is that human visual systems respond logarithmically to light energy. The results of applying the SNR measure to our denoising of *noisy Lena* are summarized in Table 2.8. Unlike our other measures, an increase in SNR represents a decrease in error. But, as with the other measures discussed so far, the SNR also does not seem to accurately quantify the amount of decrease in noise in the two denoisings of *noisy Lena.*

The measures of error discussed above have been used for many years. Their deficiencies in relation to accurate quantification of the perceptions of our visual systems are well known. It is generally recognized that they have remained in use, despite their deficiencies, mostly because they have fit well into the type of mathematics used in image processing, making theoretical predictions concerning their values relatively easy to obtain. However, because of the recent explosion of applications of wavelet analysis, it follows that we should also use some measure of error that fits well into a wavelet analysis framework. One such measure, which we shall denote by $\|\mathbf{f}-\mathbf{g}\|_{\mathrm{w}}$, is defined as follows. Suppose that $\{\widehat{f}_{j,k}\}$ are the wavelet transform values for the image $\mathbf{f}$ using one of the Daubechies wavelet transforms, and suppose that $\{\widehat{g}_{j,k}\}$ are transform values for the image $\mathbf{g}$ using the same wavelet transform. The quantity $\|\mathbf{f}-\mathbf{g}\|_{\mathrm{w}}$ is then defined by

$$\|\mathbf{f}-\mathbf{g}\|_{\mathrm{w}} = \frac{|\widehat{f}_{1,1} - \widehat{g}_{1,1}| + |\widehat{f}_{1,2} - \widehat{g}_{1,2}| + \cdots + |\widehat{f}_{N,M} - \widehat{g}_{N,M}|}{MN}.$$

For the *noisy Lena* image we used a 9-level[11] Coif12 transform to compute $\|\mathbf{f}-\mathbf{g}\|_{\mathrm{w}}$. The results are shown in Table 2.8. Notice that this new, wavelet based measure of error finally produces results that more effectively match our perceptions of the success of these denoisings. For example, for the soft thresholding denoising, this new measure shows that there is almost a 50% noise reduction. This seems a more accurate reflection of our perception of the improved quality of the denoised image, certainly more accurate than the reductions of around 10% for the other measures. Of course, one example does not prove the worth of this new measure of error. It does, however, provide a stimulus for further investigation. The reader is invited to carry out more denoisings of the noisy images found at the FAWAV website and compare the various error measures.

[11]Nine levels is the maximum number of levels possible for a 512 by 512 image.

Table 2.8 Error measurements for Figure 2.20

Image	*RMS*	$\mathcal{D}(\mathbf{f},\mathbf{g})$	*SNR*	$\|\mathbf{f}-\mathbf{g}\|_w$
noisy image	17.4	0.131	17.8	9.10
hard thresh. denoise	15.8	0.119	18.7	5.00
soft thresh. denoise	15.2	0.115	19.0	4.62

A couple more remarks need to be made about this new wavelet based measure of error. First, it is not the only measure that can be defined using wavelet transform values. For instance, the terms $|\widehat{f}_{j,k} - \widehat{g}_{j,k}|$ can be raised to powers greater than 1, and/or multiplied by weighting factors that vary depending on what level the transform values belong to. These factors could be chosen, for example, to reflect the different sensitivities of the human visual system to different fluctuation levels. Second, it is worthwhile to reflect on some of the reasons why these new wavelet based measures may provide better noise reduction estimates. Some of the deep mathematical reasons are examined in papers that are listed in this chapter's references. Besides pure mathematical reasons, it may also be that the multiresolution analysis and thresholding performed via wavelets is analogous to the threshold processing performed by the networks of neurons in the human brain in order to decompose visual images into multiple resolutions. Some theoretical models have been proposed for understanding human vision that rely on such analogies.[12]

Further aspects of noise removal

We conclude our discussion of wavelet based denoising of images by briefly outlining some other aspects of this fascinating new field. First, it is important to point out that thresholding—of either the hard or soft kind—is just one approach to denoising of images. Another approach is to modify a thresholding procedure by reducing the size of the threshold to allow in more noisy transform values and, more importantly, more image transform values. The image transform values can often be distinguished from the noise values by using the following principle: *When significant values are found at the same relative locations in each fluctuation subimage, then these values are most likely image values.* The rationale behind this principle is that the values from the image are samples of a smooth function, which will be approximated at lower resolution in each trend subimage, thus pro-

[12]Some of the papers that describe wavelet-like models for human vision are listed in the Notes and references for this chapter.

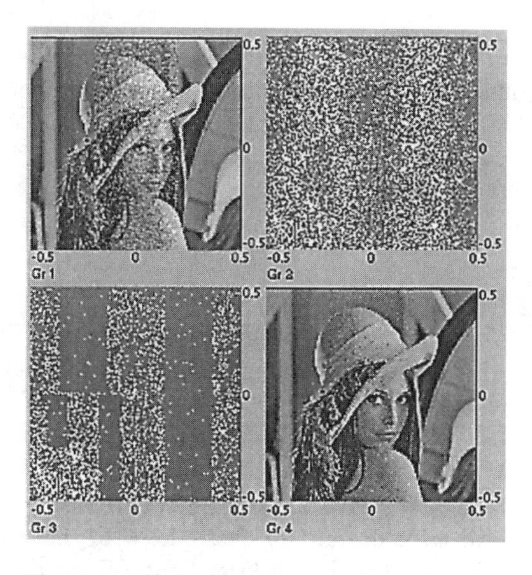

FIGURE 2.22
Removing strip noise: Gr 1, *Lena* with vertical strip noise; 4-level Coif12 transform of noisy image; Gr 3, denoised transform; Gr 4, denoised image.

ducing a hierarchy of similar fluctuation subimages along levels. This can be seen clearly in the octagon image transform in Figure 2.15(d) and in the transform of the *Barb* image in Figure 4.2(a). An application of these ideas to denoising medical images is described in the references.

Second, the fact that most wavelets are supported over small squares implies that wavelet denoising can focus on specific regions of an image where the noise is concentrated. For example, consider the noisy version of the *Lena* image shown at the top left of Figure 2.22. In this image the noise is concentrated along a vertical strip through the middle of the image. Noise analogous to this strip noise occurs, for example, in MRI imaging. Notice that the 4-level Coif12 transform of this noisy image has kept the noise concentrated in vertical strips along the middle of each fluctuation subimage. This is a consequence of the localized support of the Coif12 wavelets. By restricting the soft thresholding to just these vertical strips, and not modifying at all the regions outside these strips, it is possible to reduce the noise along the strip in the image while leaving the noise-free regions outside the strip essentially unaltered. At the lower left of Figure 2.22 we show the transform modified by soft thresholding on the strips in the first and second level fluctuations. The inverse transform of this modified transform is at the lower right of Figure 2.22. This denoised image shows some errors, but they are concentrated in the vertical strip where the noise was. Outside this vertical strip, the denoised image values are unchanged from their original, uncontaminated values. A summary of error measures is shown in Table 2.9. They indicate that this denoising was quite effective. The wavelet based measure, in particular, shows that the denoising reduced the noise by close to a factor of 3.

Table 2.9 Error measurements for Figure 2.22

Image	*RMS Error*	$\|\mathbf{f}-\mathbf{g}\|_{\mathrm{w}}$
noisy image	15.54	7.66
denoised image	8.59	2.68

As a final example, we show an example of denoising of *multiplicative noise.* This type of noise occurs, for example, with the *speckle noise* that contaminates images produced with laser light. When there is multiplicative noise, the noisy image $\mathbf{f}$ satisfies

$$\mathbf{f} = \mathbf{s}\,\mathbf{n}, \tag{2.49}$$

where $\mathbf{s}$ is the original uncontaminated image and $\mathbf{n}$ is the noise. Equation (2.49) states that the values of $\mathbf{f}$ satisfy $f_{j,k} = s_{j,k}n_{j,k}$. At the left of Figure 2.23 we show a noisy version of the *Lena* image that has been contaminated with multiplicative noise whose values are all positive. Since the values of $\mathbf{f}$ are also positive (except for a few isolated zeros), we can turn this multiplicative noise into additive noise using a logarithm. By applying a base 4 logarithm to the values of the images, we obtain

$$\log_4 \mathbf{f} = \log_4 \mathbf{s} + \log_4 \mathbf{n}. \tag{2.50}$$

Actually, to be precise, a tiny fudge term of .0000001 was added to the values of $\mathbf{f}$ to prevent any logarithms of zero values.

Equation (2.50) shows that the multiplicative noise $\mathbf{n}$ has become an additive noise $\log_4 \mathbf{n}$. Applying a threshold denoising to the image $\log_4 \mathbf{f}$, followed by an application of base 4 exponentiation, we obtained the denoised image shown at the right of Figure 2.23. The summary of error measures shown in Table 2.10 reveals how effective this denoising was. The wavelet based measure, for instance, shows that the noise was reduced by a factor of 3.

We have tried in this section to indicate several of the many features that wavelet based denoising enjoys. Our apologies to the reader if he or she is tired of seeing *Lena*. We concentrated on this particular image in order to reduce the number of figures. At the FAWAV website, we provide many more images on which the reader may test these ideas.

FIGURE 2.23
Removing multiplicative noise: Gr 1, *Lena* with multiplicative random noise; Gr 2, denoised image.

Table 2.10 Error measurements for Figure 2.23

Image	*RMS Error*	$\|\mathbf{f} - \mathbf{g}\|_w$
noisy image	20.21	14.73
denoised image	10.08	4.70

2.11 Some topics in image processing

We conclude our introduction to 2D wavelet analysis with a brief discussion of a few examples of wavelet based techniques in image processing. These examples are meant to provide the reader with a small sampling from the huge variety of applications that have appeared in the last few years. Further examples can be found in the references for this chapter.

Edge detection

One task in image processing is to produce outlines—the edges—of the various objects that compose an image. As shown in Figure 2.15, a wavelet transform can detect edges by producing significant values in fluctuation subimages. One way to produce an image of these edges is to simply perform an inverse transform of only the fluctuation values, after setting all the trend values equal to zero. This produces an image of the first-level detail image $\mathbf{D}^1$. For example, in Figure 2.24(a) we show such a modification of the 1-level Coif6 transform of the octagon image obtained by setting all its first trend values equal to zero. By performing an inverse transform of

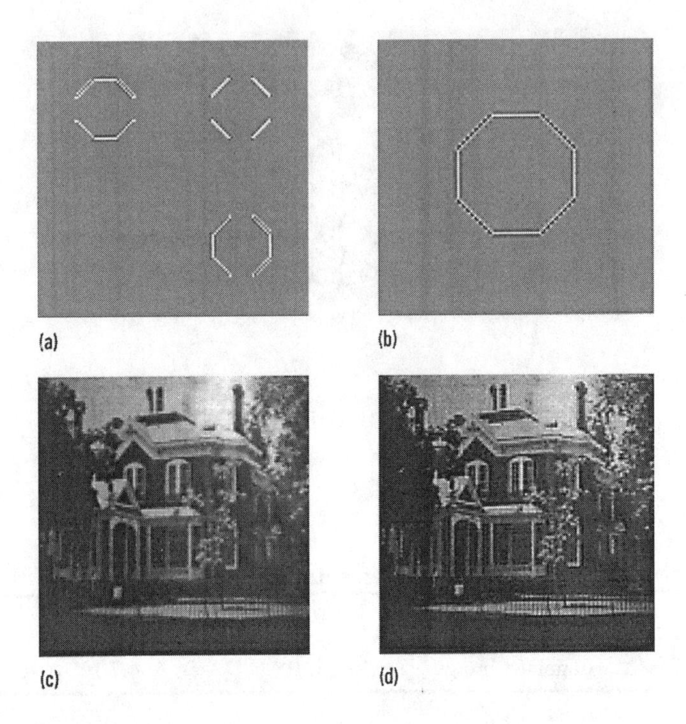

FIGURE 2.24
(a) 1-level Coif6 transform of octagon image with trend subimage values all replaced by zeros. (b) Inverse transform of image in (a). (c) Image of a house. (d) Edge enhanced image.

this modified transform, we produced the detail image $\mathbf{D}^1$ shown in Figure 2.24(b). This figure clearly highlights the edges of the original image. Of course, this image could be further processed to increase the highlighting of just the outline of the octagon. For instance, only the most intense, whitest portion of the image could be retained, using a thresholding operation.

Edge enhancement

If we can detect edges accurately, then we can also enhance their appearance in an image. This will sharpen images that suffer from dull, blurry edges. For example, in Figure 2.24(c) we show an image of a house that suffers from blurred edges. In order to sharpen the edges of this image, we used the following method.

Edge Enhancement Method

Step 1. Perform a wavelet transform of image.

Step 2. Multiply fluctuation values by a constant larger than 1, and leave trend values unchanged.

Step 3. Perform an inverse transform of modified image from Step 2.

To produce the edge enhanced image of the house, we used a 1-level Daub4 transform in Step 1, and we multiplied the first-level fluctuation values by the constant 3 in Step 2.

Comparing the two images of the house in Figure 2.24, we can see that the edge enhanced image is a sharper image than the original. Some details can be seen more clearly in the enhanced image, such as the fencing in front of the house, and the woodwork at the top of the roof. Particularly striking is the fact that in the right central window a vertical window divider has been rendered visible in the enhanced image. This kind of edge sharpening and improvement in detail visibility is clearly of fundamental importance in medical and biological imaging. The edge enhancement method just described is only the simplest of many related procedures. For instance, if a multiple level transform is performed in Step 1, then Step 2 can be done differently by using different constants for each fluctuation level.

Image recognition

Teaching a machine to recognize images is a very difficult problem. For example, we might want to program a computer to recognize images of human faces. Image recognition is important in areas such as data retrieval, object identification, and the science of human vision. *Data retrieval* refers to finding a match to a given image from a huge archive of images and then retrieving more data associated with that image (e.g., name, date of birth, etc.). *Object identification* refers to the problem of identifying the location, or lack thereof, of a particular object such as an aircraft within a complicated scene. Image retrieval relates to some aspects of the *science of human vision* because of our desire to know how our brains recognize images out of the data supplied by our eyes. While programming a computer to recognize an image is not the same as understanding how our brains do it, nevertheless, insights gained from the one task will certainly provide ideas for the other.

In Figure 2.25 we illustrate an elementary problem of image recognition. Suppose that we want a machine to identify which of the four images shown in Figure 2.25(a) is the best match to the denoised version of *Lena* in Figure 2.20(a). We choose to look for a match to the *denoised Lena,* rather than *Lena,* since it is more likely that we would not receive a precise image for matching, but rather a noisy version. It is also likely that the image we wish to match differs in other ways (different vantage point, different facial expression, etc.) from the image of that person in the archive. This raises the considerably more difficult problem of matching distinct images

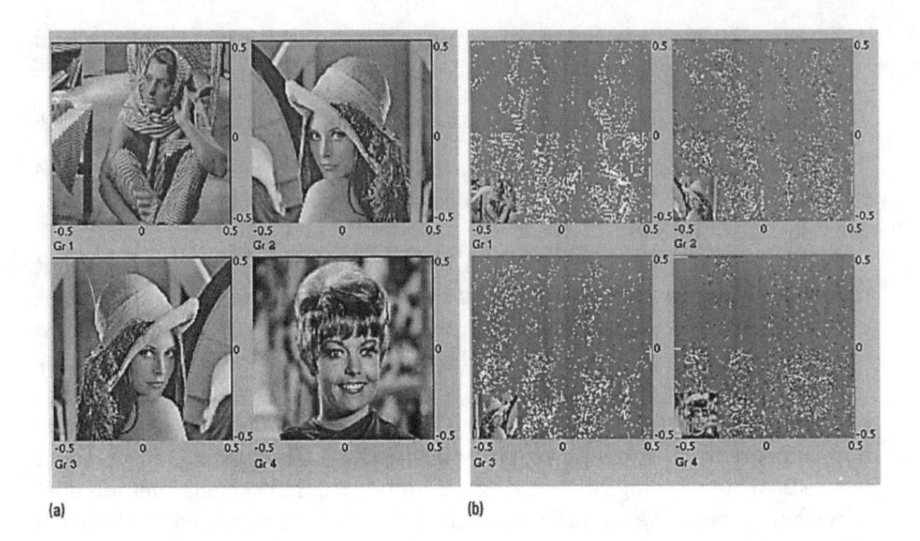

FIGURE 2.25
(a) Four images. (b) 2-level Coif18 transforms of these images.

of the same person. It turns out that the wavelet based algorithm for image matching outlined below helps with this problem, too.

A simple method for matching would be to compute an error measure, say $\mathcal{D}(\mathbf{f}, \mathbf{g})$, where $\mathbf{f}$ is the denoised *Lena* image and $\mathbf{g}$ is each of the images in our archive. For the four images in Figure 2.25(a) this method works well; as shown in the column labeled *Full* in Table 2.11, the image labeled Gr 3 produces $\mathcal{D}(\mathbf{f}, \mathbf{g}) = 0.066$, and all other images produce $\mathcal{D}(\mathbf{f}, \mathbf{g}) > 0.48$. So the image in Gr 3 is clearly the best match. There is a significant problem with this approach, however, since an actual archive will certainly contain far more than four images—it might contain tens of thousands of images. The images we are using are 512 by 512, hence each image contains over a quarter of a million values. The time delay—due to the enormous number of calculations—associated with finding $\mathcal{D}(\mathbf{f}, \mathbf{g})$ for tens of thousands of images is prohibitively large.

Wavelet analysis provides a flexible approach for significantly reducing the number of computations needed for this image matching problem. The wavelet based approach also incorporates its own compression procedure, which allows for the archive to be stored in a compressed format. To illustrate this wavelet approach to image matching, consider the 2-level Coif18 transforms of the four images in our small archive shown in Figure 2.25(b). Notice that our visual systems can easily distinguish the four images based only on their second trend subimages, which are one-sixteenth the size of the original image. These smaller images are called *thumbnail* images. If we simply compute the error measures for these second trends only, then we obtain the results shown in the column labeled *Second* in Table 2.11. This new computation easily matches the *denoised Lena* image with the correct image of *Lena* in our tiny archive. Since the second trends are one-sixteenth the size of the original images, this new computation is sixteen times faster than the first one involving the full images.

Table 2.11 Error measures for images

Image	*Full*	*Second*	*Third*	*Fourth*	*Fifth*
Gr 1	0.481	0.495	0.473	0.459	0.426
Gr 2	0.485	0.491	0.475	0.457	0.410
Gr 3	0.066	0.076	0.063	0.063	0.063
Gr 4	0.546	0.549	0.540	0.529	0.501

What works well for second-level trends also works well for third, fourth, and even fifth trends. As shown in Table 2.11, computing errors for any of these trends clearly determines the correct match. If we use fifth trends to perform error computations, then our calculations proceed over 1000 times faster than if we were to use the full images. This improved performance factor makes it quite practical to search through an archive of tens of thousands of images. It is important to note that in order to achieve this rapid performance, the images must be stored in the archive in their wavelet transformed format. Fortunately, this also allows for the images to be stored in a compressed format.

Another advantage of comparing these very low resolution, fifth trend images is the following. If the given image is only a fair approximation, rather than a perfect replica, of its version in the archive, then the low resolution fifth trend versions should still produce smaller error measures than the fifth trends of very different looking people. This indicates how a wavelet based algorithm facilitates the matching of distinct images of the same person.

Of course, if we are using fifth trends it is quite possible that several images of different people might produce errors that are less than whatever error threshold we have established for matching. In that case, by using the fourth-level fluctuations, we can rapidly construct the fourth-level trend subimages for the subset of images which survived the fifth-level matching test. The error measures can then be computed for these fourth-level trends. Or, better yet, we could instead bypass the reconstruction of the fourth-level trends, and for each image add on an error term computed from the fourth-level fluctuations to the error already computed for the fifth-level trend. This is equivalent to computing errors for the fourth-level trends, and is a faster computation. Although these additional calculations require more time, it is very likely that the subset of images that survived the fifth trend comparison is much less numerous than the original set. The recursive structure of wavelet transforms enables the implementation of a recursive matching algorithm that successively winnows the subset of possible matching images, using extremely small numbers of computations to distinguish grossly dissimilar images, while reserving the most numerous computations only for a very small number of quite similar images.

This brief outline of an image retrieval algorithm should indicate some of the fundamental advantages that a wavelet based approach provides. One of the most important features of the approach we described is its use of the multiresolution structure (the recursive structure of the trend subimages) to facilitate computations. This has applications to many other problems. For instance, in object identification, a method known as *correlation,* which we will describe in the next chapter, is a computationally intensive method for locating a given object within a complicated scene. Using correlation on third trends instead of the full images, however, can greatly speed up the approximate location of an object.

2.12 Notes and references

Descriptions of the many other types of wavelets—biorthogonal wavelets, spline wavelets, etc.—can be found in [BGG], [DAU], [MAL], [REW], or [STN]. The book [CHU] is particularly good on the topic of spline wavelets. Two good references on digital image processing are [JAH] and [RUS]. The book [ALU] contains several excellent papers on wavelets and their applications in biology and medicine.

Further discussions of quantization can be obtained from [MAL], [VEK], or [DVN]. Some good references on information theory are [ASH], [COT], and [HHJ].

An excellent summary of wavelet-based image compression can be found in [DVN]. A good site for downloading images is located at:

`http://links.uwaterloo.ca/bragzone.base.html`

Details on the best image compression ratios yet obtained can be viewed at the following website:

`http://www.icsl.ucla.edu/~ipl/psnr_results.html.`

A basic image compression software package can be downloaded from the website:

`http://www.cs.dartmouth.edu/~gdavis/wavelet/wavelet.html`.

This site also has a number of useful links to other sites.

The zero-tree method of image compression is described in the papers [SHA] and [SAP]. Software that implements the zero-tree method can be obtained by FTP from the following address:

`ftp://ipl.rpi.pub/EW_Code`

or can be downloaded from the following website:

`http://www.ipl.rpi.edu/SPIHT`.

There is an interesting analysis of tree structures in wavelet-based image compression and their relationship with fractal image compression in [DVS].

Formulas for the RGB to IHS mapping can be found in [RUS]. The differences in sensitivity of the human visual system to I versus H and S can be found in [WAN]. The book [WAN] is the best elementary introduction to the science of the human visual system. In [WAN] there is a good summary of MRA and its relation to vision. Wavelet-like models of the human visual system are described in [WAT], [FI1], and [FI2].

The best discussions of fingerprint compression can be found in the articles [BBH] and [BRS] by the creators of the WSQ method. The website:

`http://www.c3.lanl.gov/~brislawn`

may also be of interest. One of the fingerprints, Fingerprint 8, at the FAWAV website is a cropped version of a fingerprint found at this site. It was supplied to the author courtesy of Dr. Brislawn. The other fingerprints, including Fingerprint 1 in Figure 2.18(a), were obtained from the NIST database at the ftp site:

`sequoyah.nist.gov`.

Introductory treatments of noise can be found in [BAS] and [FAN]. The method of choosing denoising thresholds for wavelet transforms has received a very complete analysis in the papers [DOJ] and [DJK], and is given a good summary in [MAL]. There is also an excellent survey of various wavelet-based noise removal techniques in [DON]. The use of different thresholds for each fluctuation subimage is utilized in the program SURESHRINK, which can be downloaded from the following website:

`http://www.cs.scarolina.edu/ABOUT_US`

`/Faculty/Hilton/shrink-demo.html`.

Another commonly used measure of errors in images is PSNR, which is closely related to SNR; see [VEK] for the definition of PSNR. A discussion of new wavelet-based measures of errors in images is given in [CDL]. Removing noise in medical images, based on thresholding and based on the statistics of significant values between levels, is described in [MLF] and [XWH].

There is an interesting discussion in [WI2] of a wavelet based solution of the *rogues gallery problem,* the problem of retrieving an image from a large archive. This discussion differs somewhat from the one given here.

Chapter 3

Frequency analysis

> If our ear uses a certain technique to analyze a signal, then if you use that same mathematical technique, you will be doing something like our ear. You might miss important things, but you would miss things that our ear would miss too.
>
> *Ingrid Daubechies*[1]

It is well known that human hearing responds to the frequencies of sound. Evolution has shaped our sense of hearing into a superb frequency analyzer, whose full characteristics we do not yet completely understand. Nevertheless, it is clear that we perceive tones of different frequency as having different pitch, and musical notes are called higher or lower depending on whether they have a corresponding higher or lower frequency.

The mathematical analysis of the frequency content of signals is called *Fourier analysis.* In this chapter we shall sketch the basic outlines of Fourier analysis as it applies to discrete signals and use it to analyze the frequency content of wavelets. A deeper understanding of wavelets can be gained from studying their frequency content, and by examining how this frequency content relates to wavelet transforms of signals. To keep the mathematics as simple as possible we shall focus on 1D signals, although in Section 3.5 we shall describe some 2D applications.

3.1 Discrete Fourier analysis

The frequency content of a discrete signal is revealed by its *discrete Fourier transform* (DFT). The DFT of a discrete signal is usually per-

[1]Daubechies' quote is from [BUR].

formed via a computer using an algorithm called a *fast Fourier transform* (FFT). No attempt will be made in this primer to prove any of the results about the DFT that we shall be using. The emphasis instead will be on how the DFT is applied in signal analysis, especially in wavelet analysis. There are many excellent treatments of the DFT and the FFT available for the reader who desires further discussion and proofs; we list a few of these references at the end of this chapter. In any case, the results we shall be using are all standard ones that are quite well known. Our approach will be to illustrate some fundamental ideas via a few examples rather than via a theoretical derivation.

The DFT reveals the frequency content of a discrete signal; so we shall begin by reviewing the notion of frequency. The analog signals $\cos 2\pi\nu x$ and $\sin 2\pi\nu x$, where x denotes time, both have a fundamental time period of $1/\nu$. Consequently, these signals repeat their basic form ν times in a unit-length time interval, i.e., they have a *frequency* of ν cycles/unit-time. The notion of frequency of a discrete signal is closely related to frequency of analog signals, as the following examples show. In the next section we shall make this connection more precise, but it may help to first examine some illustrative examples.

As a first example of a DFT, consider the discrete signal, Signal I, shown in Figure 3.1(a). Signal I consists of 1024 samples of the analog signal

$$g_1(x) = 2\cos 4\pi x + 0.5 \sin 24\pi x. \tag{3.1}$$

It has frequencies of 2 and 12 for its two terms. In Figure 3.1(b) we show the DFT of Signal I. The two inner spikes are located at ± 2, corresponding to the frequency 2 of the cosine term of Signal I, and the two outer spikes are located at ± 12, corresponding to the frequency 12 of the sine term of Signal I. Because of the way in which the DFT is defined (see the next section), the spikes appear at $\pm\nu$ instead of just at ν when the signal contains a sine or cosine term of frequency ν. Notice that the lengths of the spikes in the DFT of Signal I that correspond to the cosine term in Signal I appear to be about 4 times greater than the lengths of the spikes that correspond to the sine term, and 4 is also the ratio of the amplitude of the cosine term to the amplitude of the sine term in Signal I. This example shows how the DFT can identify the frequencies—and the relative sizes of their contributions in terms of amplitude—that make up the frequency content of a signal.

As a second example of a DFT, consider the discrete signal, Signal II, shown in Figure 3.1(c). Signal II consists of 1024 samples of the analog signal

$$g_2(x) = \frac{1 + \cos 24\pi x}{1 + 4x^2}. \tag{3.2}$$

Its DFT is shown in Figure 3.1(d). The significant values in the DFT are clustered around 0 and ± 12. These values of ± 12 clearly correspond to

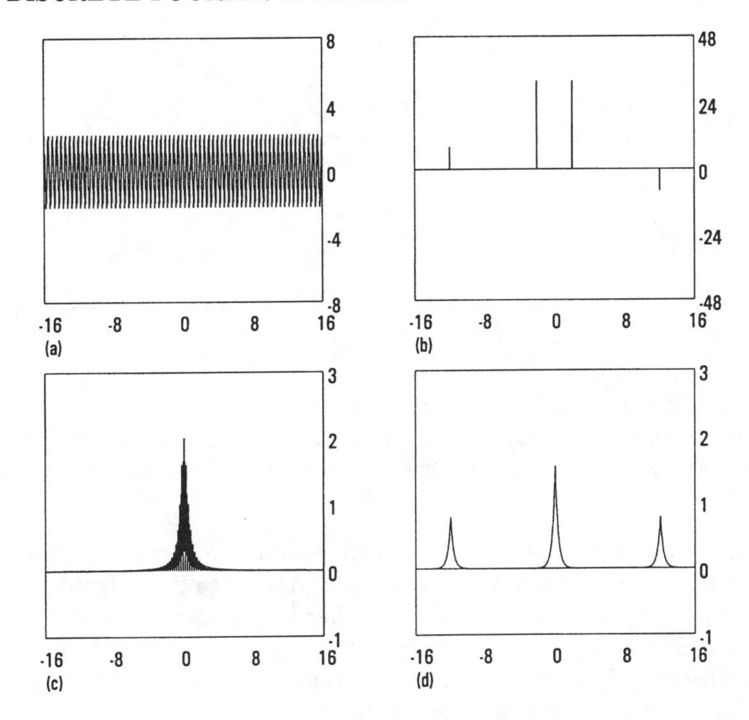

FIGURE 3.1
(a) Signal I. (b) DFT of Signal I. (c) Signal II. (d) DFT of Signal II.

the frequency 12 in the cosine term in Formula (3.2), but the cause of the cluster of significant DFT values around 0 is less clear. We shall see what these very low frequency values have to do with Signal II in Section 3.3, when we examine the frequency content of averaged signals in wavelet multiresolution analysis.

Frequency content of wavelets

As a third example of DFTs, we consider the frequency content of scaling signals and wavelets. This will enable us to comprehend the effects of wavelet MRAs on the frequency content of a signal, which we shall examine in Section 3.3.

As a typical case of scaling signals and wavelets, consider the Coif12 scaling signal $\mathbf{V}_1^1$ and wavelet $\mathbf{W}_1^1$. In Figure 3.2(a) we show the squares of the magnitudes of the DFT of the Coif12 scaling signal $\mathbf{V}_1^1$, which is called the *spectrum* of this signal. Notice that the significant, non-zero values of this spectrum are concentrated in the central half of the graph near the origin, which are the lower frequencies. The higher frequencies lie in the left and right quarters of the graph, and for these higher frequencies the

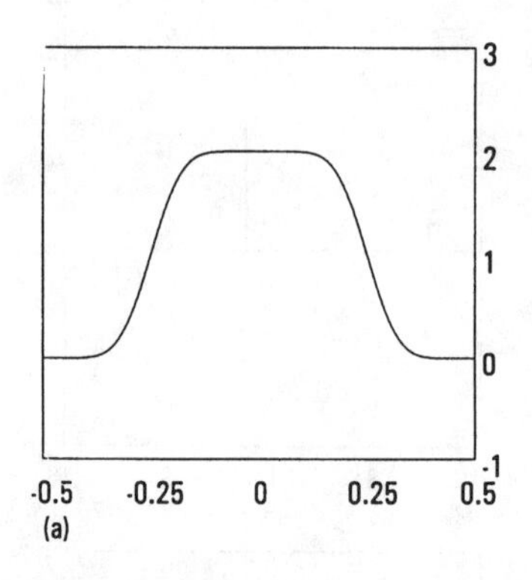

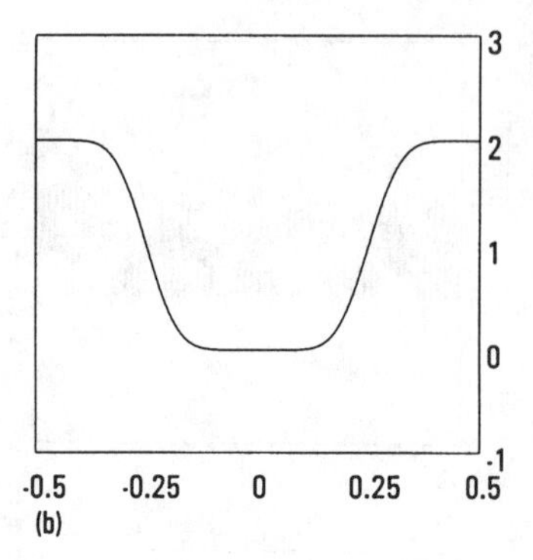

FIGURE 3.2
(a) Squares of magnitudes of DFT of Coif12 scaling signal $\mathbf{V}_1^1$, its spectrum. (b) Spectrum of Coif12 wavelet $\mathbf{W}_1^1$.

values of the spectrum are approximately zero. In contrast, consider the spectrum of the wavelet $\mathbf{W}_1^1$, shown in Figure 3.2(b). Its significant, non-zero values are concentrated in the higher frequencies, while for the lower frequencies its values are approximately zero.

Notice that the graphs in Figure 3.2 seem complementary to one another. In fact, it is the case that the sum of their values equals the constant 2. This is one of the fundamental properties of the spectrums of first-level scaling signals and wavelets.

3.2 Definition of the DFT and its properties

In this section, we shall state the definition of the DFT and some of its principal properties. We shall be fairly brief; more thorough discussions can be found in many references, some of which are listed at the end of this chapter.

To state the definition of the DFT in a succinct form, we shall make use of the Σ-notation for summation. A sum $g_1 + g_2 + \cdots + g_M$ can be more compactly expressed as

$$\sum_{m=1}^{M} g_m.$$

For instance, the following sum

$$a + ar + ar^2 + \cdots + ar^{M-1} = \sum_{m=1}^{M} ar^{m-1}$$

should be familiar from algebra.

Using this summation notation, we can state the definition of the DFT of a signal $\mathbf{f}$ of length N. We shall denote this DFT by $\mathcal{F}\mathbf{f}$, and its values $(\mathcal{F}\mathbf{f})_n$ are defined by

$$(\mathcal{F}\mathbf{f})_n = \sum_{m=1}^{N} f_m e^{-i2\pi(n-1)(m-1)/N}. \tag{3.3}$$

Although the variable n in (3.3) can be any integer, we shall see that the periodicity property stated below implies that the values $(\mathcal{F}\mathbf{f})_n$ for $n = -N/2$ to $n = N/2 - 1$ are sufficient for describing $\mathcal{F}\mathbf{f}$.

In Formula (3.3) we use the complex exponentials $e^{-i2\pi(n-1)(m-1)/N}$, which are defined via *Euler's formulas:*

$$e^{i\theta} = \cos\theta + i\sin\theta \tag{3.4}$$

and

$$e^{-i\theta} = \cos\theta - i\sin\theta. \tag{3.5}$$

One consequence of (3.5) is that the modulus (or magnitude), $|e^{-i\theta}|$, of the complex number $e^{-i\theta}$ is equal to 1. That is, $e^{-i\theta}$ lies on the unit-circle in the complex plane.

Another consequence of (3.4) and (3.5) is that

$$\cos 2\pi\nu x = \frac{1}{2}e^{-i2\pi\nu x} + \frac{1}{2}e^{i2\pi\nu x}. \tag{3.6}$$

This equation shows that $\cos 2\pi\nu x$ can be expressed as a sum of two complex exponentials having frequencies $\pm\nu$, which should remind the reader of the DFT of Signal I discussed in the previous section. Similarly,

$$\sin 2\pi\nu x = \frac{i}{2}e^{-i2\pi\nu x} - \frac{i}{2}e^{i2\pi\nu x} \tag{3.7}$$

which shows that $\sin 2\pi\nu x$ can also be expressed as a sum of complex exponentials having frequencies of $\pm\nu$.

The following four properties are the principal ones satisfied by the DFT. We use the notation $\mathbf{f} \overset{\mathcal{F}}{\longmapsto} \mathcal{F}\mathbf{f}$ to symbolize the DFT operation.

Properties of the DFT

1. *Linearity.* For all constants α and β, and all signals $\mathbf{f}$ and $\mathbf{g}$ of length N,

$$\alpha\mathbf{f} + \beta\mathbf{g} \overset{\mathcal{F}}{\longmapsto} \alpha\mathcal{F}\mathbf{f} + \beta\mathcal{F}\mathbf{g}.$$

2. *Periodicity.* If $\mathbf{f}$ is a signal of length N, then $\mathcal{F}\mathbf{f}$ has period N; that is,

$$(\mathcal{F}\mathbf{f})_{n+N} = (\mathcal{F}\mathbf{f})_n \tag{3.8}$$

holds for all integers n.

3. *Inversion.* The signal $\mathbf{f}$ can be obtained from $\mathcal{F}\mathbf{f}$ by

$$f_m = \frac{1}{N} \sum_{n=-N/2}^{N/2-1} (\mathcal{F}\mathbf{f})_n e^{i2\pi(n-1)(m-1)/N} \tag{3.9}$$

for $m = 1, 2, \ldots, N$.

4. *Parseval's Equality.* The signal $\mathbf{f}$ and its DFT $\mathcal{F}\mathbf{f}$ satisfy

$$\sum_{m=1}^{N} |f_m|^2 = \frac{1}{N} \sum_{n=-N/2}^{N/2-1} |(\mathcal{F}\mathbf{f})_n|^2 . \tag{3.10}$$

Because of periodicity, the N values $(\mathcal{F}\mathbf{f})_n$ for $n = -N/2$ to $n = N/2 - 1$ are sufficient for uniquely determining a DFT $\mathcal{F}\mathbf{f}$. It is for this reason that all of our figures of DFTs have the same number of values as the signal being transformed; and all software packages that compute DFTs follow this same convention.

The Inversion Formula (3.9) is particularly important. For one thing, it ensures that distinct signals must have distinct DFTs. For another, it allows for a useful interpretation of DFT values. Each DFT value $(\mathcal{F}\mathbf{f})_n$ is an amplitude for a discrete complex exponential signal

$$e^{i2\pi(n-1)(m-1)/N}, \quad m = 1, 2, \ldots, N, \tag{3.11}$$

which is a sampled version of $e^{i2\pi(n-1)x}$, a complex exponential analog signal of frequency $n - 1$. The sample points are $x_m = (m-1)/N$. Or, since

$$\frac{(n-1)(m-1)}{N} = \frac{n-1}{\Omega} \cdot \frac{(m-1)\Omega}{N},$$

one can also view (3.11) as sample values of $e^{i2\pi(n-1)x/\Omega}$, a complex exponential of frequency $(n-1)/\Omega$. In this latter case, the sample points are $x_m = (m-1)\Omega/N$. This latter case is important when signals are obtained from measured samples of analog signals over a time interval of length Ω. In any case, the Inversion Formula (3.9) shows that the signal $\mathbf{f}$ can be realized by summing these discrete complex exponential signals with amplitudes given by the DFT values $(\mathcal{F}\mathbf{f})_n$, and multiplying by the scale factor $1/N$.

Parseval's Equality (3.10) can be regarded as a Conservation of Energy property, provided we include a scale factor of $1/N$ again. This Conservation of Energy, as with wavelet transforms, is useful for understanding how to make applications of the DFT operation. In fact, in some instances, the DFT, or transforms closely related to it, can be used for compression and noise removal. While these are fascinating topics, there is insufficient space in this primer for discussing it any further. We do give some references, however, in the last section of this chapter.

Another way of interpreting Parseval's Equality is to observe that the left side of (3.10) is equal to the energy $\mathcal{E}_{\mathbf{f}}$ of the signal $\mathbf{f}$, and that the right side of (3.10) is equal to the average, or mean, value of the spectrum $|\mathcal{F}\mathbf{f}|^2$ of $\mathbf{f}$. That is, Parseval's Equality states that *the energy of a signal* $\mathbf{f}$ *is equal to the mean value of its spectrum* $|\mathcal{F}\mathbf{f}|^2$.

One final remark needs to be made about periodicity and inversion. They imply, since the right side of (3.9) also has period N in the integer variable m, that the finite signal $\mathbf{f}$ should be assumed to be a subsignal of a periodic signal having period N. That is,

$$f_{m+N} = f_m \tag{3.12}$$

for all integers m. Notice that (3.12) is the wrap-around property of signals that we made use of for scaling signals and wavelets in Chapter 2.

z-transforms *

The z-transform provides a more flexible way of expressing the values of the DFT, which is especially helpful in wavelet theory. Since some readers may find the theory of z-transforms to be difficult, we shall treat it as optional material. This material will be used later only in optional sections.

The z-transform of $\mathbf{f}$ will be denoted by $\mathbf{f}[z]$ and is defined by

$$\mathbf{f}[z] = \sum_{m=1}^{N} f_m z^{m-1}. \tag{3.13}$$

The variable z takes its values on the unit-circle of the complex plane.[2]

If we set z equal to $e^{-i2\pi(n-1)/N}$, then

$$\mathbf{f}[e^{-i2\pi(n-1)/N}] = (\mathcal{F}\mathbf{f})_n. \tag{3.14}$$

Formula (3.14) shows that the DFT, $\mathcal{F}\mathbf{f}$, consists of the values of $\mathbf{f}[z]$ at the points $z = e^{-i2\pi(n-1)/N}$ which lie uniformly spaced around the unit-circle in the complex plane.

One application of the z-transform is to the computation of DFTs of scaling signals and wavelets. To do this, we must first define the *cyclic translation* of a signal $\mathbf{f}$. The cyclic translation forward by 1 unit is denoted by $\mathcal{T}_1\mathbf{f}$ and is defined by

$$\mathcal{T}_1\mathbf{f} = (f_N, f_1, f_2, \ldots, f_{N-1}). \tag{3.15}$$

[2]There is another definition of z-transform that is also used in signal processing. The one we are working with is most useful for finite length signals.

Notice the wrap-around at the end of $\mathcal{T}_1\mathbf{f}$; because of this wrap-around we use the adjective *cyclic* in describing this translation. The cyclic translation backward by 1 unit is denoted by $\mathcal{T}_{-1}\mathbf{f}$ and is defined by

$$\mathcal{T}_{-1}\mathbf{f} = (f_2, f_3, \ldots, f_N, f_1). \tag{3.16}$$

All other cyclic translations are defined via compositions, e.g., $\mathcal{T}_2 = \mathcal{T}_1 \circ \mathcal{T}_1$, $\mathcal{T}_3 = \mathcal{T}_1 \circ \mathcal{T}_1 \circ \mathcal{T}_1$, and so on. The most important property of these cyclic translations is that $\mathcal{T}_k \circ \mathcal{T}_m = \mathcal{T}_{k+m}$ for all integers k and m.

The key property of the z-transform is that

$$\mathcal{T}_k\mathbf{f}[z] = \mathbf{f}[z]z^k \tag{3.17}$$

which holds *provided z is equal to $e^{-i2\pi(n-1)/N}$ for some integer n.* To see why (3.17) holds, we demonstrate it for $k = 1$, and higher powers of k follow by repeating the argument for $k = 1$. If $k = 1$, then

$$\mathcal{T}_1\mathbf{f}[z] = f_N + f_1 z + \cdots + f_{N-1}z^{N-1}.$$

But, if $z = e^{-i2\pi(n-1)/N}$, then $z^N = 1$. Consequently

$$\begin{aligned}\mathcal{T}_1\mathbf{f}[z] &= f_1 z + \cdots + f_{N-1}z^{N-1} + f_N z^N \\ &= (f_1 + f_2 z + \cdots + f_N z^{N-1})z \\ &= \mathbf{f}[z]z\end{aligned}$$

which proves (3.17) for $k = 1$.

One application of (3.17) is to the relationship between the frequency content of $\mathbf{V}_1^1$ and all the other first-level scaling signals. Since $\mathbf{V}_m^1 = \mathcal{T}_{2m-2}\mathbf{V}_1^1$, it follows from (3.17) that

$$\mathbf{V}_m^1[z] = \mathbf{V}_1^1[z]z^{2m-2}. \tag{3.18}$$

Similarly, we have

$$\mathbf{W}_m^1[z] = \mathbf{W}_1^1[z]z^{2m-2} \tag{3.19}$$

for the first-level wavelets. These last two formulas elucidate the relationship between the frequency contents of scaling signals and wavelets because of Formula (3.14), which relates z-transforms to DFTs. For example, because $|z| = 1$, it follows that

$$|\mathcal{F}\mathbf{V}_m^1|^2 = |\mathcal{F}\mathbf{V}_1^1|^2, \quad |\mathcal{F}\mathbf{W}_m^1|^2 = |\mathcal{F}\mathbf{W}_1^1|^2$$

which shows that the spectrum of each first-level scaling signal is equal to the spectrum of $\mathbf{V}_1^1$ and the spectrum of each first-level wavelet is equal to the spectrum of $\mathbf{W}_1^1$. We shall make further use of Formulas (3.18) and (3.19) in the next section.

3.3 Frequency description of wavelet analysis

In this section we shall examine how the frequency content of a signal is changed when the signal undergoes a wavelet analysis. To be more precise, we shall compare the frequency contents of the averaged signals $\mathbf{A}^k$ and the detail signals $\mathbf{D}^k$ created in a k-level MRA with the frequency content of the original signal $\mathbf{f}$.

Let's examine a 1-level Coif12 MRA of Signal II, shown in Figure 3.1(c). This example illustrates the fundamental aspects of the effect on frequency content of an MRA using a Daubechies wavelet. At the end of this section we shall discuss the mathematical details for those readers who are interested in them.

When a first averaged signal $\mathbf{A}^1$ is created, it consists of a sum of multiples of the first level scaling signals [see Formula (2.13b)]. In Figure 3.3(a) we show the first averaged signal, using Coif12 scaling signals, for Signal II. The DFT of this first averaged signal is shown in Figure 3.3(b). It is interesting to compare this DFT with the DFT of Signal II in Figure 3.1(d) and the spectrum of the Coif12 scaling signal $\mathbf{V}_1^1$ in Figure 3.2(a). In order to make this comparison, the values of the spectrum must be graphed over $[-16, 16]$ instead of $[-0.5, 0.5]$, but this can be easily done by a change of scale of the horizontal (frequency) axis. The values of this spectrum are approximately equal to the constant 2 near the origin (for the lower frequencies), and are approximately equal to the constant 0 at the left and right (for the higher frequencies).

The relation between the DFT of $\mathbf{A}^1$ and the spectrum $|\mathcal{F}\mathbf{V}_1^1|^2$ of $\mathbf{V}_1^1$ is

$$\mathcal{F}\mathbf{A}^1 \approx \frac{1}{2} \left|\mathcal{F}\mathbf{V}_1^1\right|^2 \mathcal{F}\mathbf{f}. \tag{3.20}$$

On the right side of (3.20) is the product of the two signals $|\mathcal{F}\mathbf{V}_1^1|^2$ and $\mathcal{F}\mathbf{f}$; hence this approximation says that each value $(\mathcal{F}\mathbf{A}^1)_n$ of the DFT of $\mathbf{A}^1$ is approximately equal to 1/2 times $|(\mathcal{F}\mathbf{V}_1^1)_n|^2 (\mathcal{F}\mathbf{f})_n$. Thus 1/2 times the spectrum of $\mathbf{V}_1^1$ acts as a *low-pass filter* on the values of the DFT of $\mathbf{f}$, allowing through only the low frequency values (since $|\mathcal{F}\mathbf{V}_1^1|^2 \approx 0$ for high frequency values).

It is interesting to examine the way in which this low-pass filtering affects the averaged signal. By comparing the averaged signal $\mathbf{A}^1$ in Figure 3.3(a) with the original signal $\mathbf{f}$ in Figure 3.1(c) we can see that there is much less rapid oscillation in the averaged signal; this is due to the suppression of the high frequencies by the low-pass filtering.

In contrast to the scaling signal, 1/2 times the spectrum of the Coif12 wavelet $\mathbf{W}_1^1$ acts as a *high-pass filter*, allowing through only the high frequency portions of the DFT of $\mathbf{f}$. In fact, the following approximation

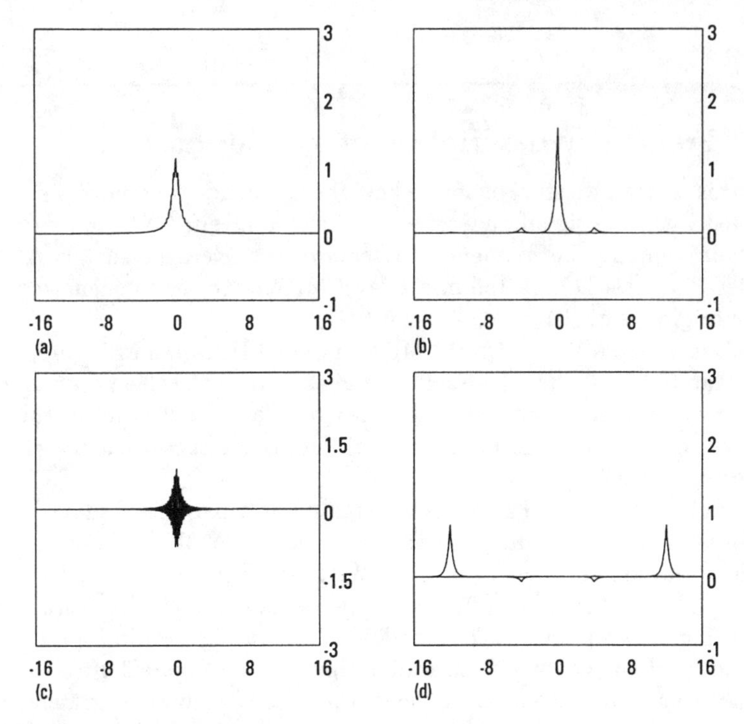

FIGURE 3.3
Frequency decomposition of 1-level Coif12 MRA. (a) First averaged signal for Signal II. (b) DFT of first averaged signal [c.f., Figures 3.1(d) and 3.2(a)]. (c) First detail signal for Signal II. (d) DFT of first detail signal [c.f., Figures 3.1(d) and 3.2(b)].

holds

$$\mathcal{F}\mathbf{D}^1 \approx \frac{1}{2}\left|\mathcal{F}\mathbf{W}_1^1\right|^2 \mathcal{F}\mathbf{f}. \tag{3.21}$$

As can be seen from Figure 3.2(a), the factor $\left|\mathcal{F}\mathbf{W}_1^1\right|^2$ is approximately zero for the low frequency values, and is approximately 2 for the high frequency values. It then follows from (3.21)—and we can check it by examining Figure 3.3(d)—that the DFT of the first detail signal mostly consists of the higher frequency values of the DFT of the signal $\mathbf{f}$. The effect that this has on the detail signal $\mathbf{D}^1$ is that it contains the most rapid, high frequency oscillations from the original signal.

This example of a 1-level Coif12 MRA of Signal II is typical of all wavelet MRAs. The frequency content of the first averaged signal $\mathbf{A}^1$ consists of low frequency values resulting from a low-pass filtering of the frequency content of the signal by 1/2 times the spectrum of the 1-level scaling signal $\mathbf{V}_1^1$. And the first-level detail signal $\mathbf{D}^1$ has a frequency content obtained from a high-pass filtering of the frequency content of the signal by 1/2 times

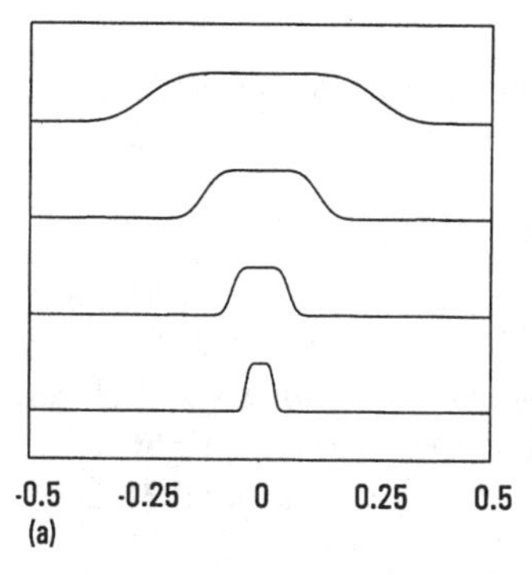

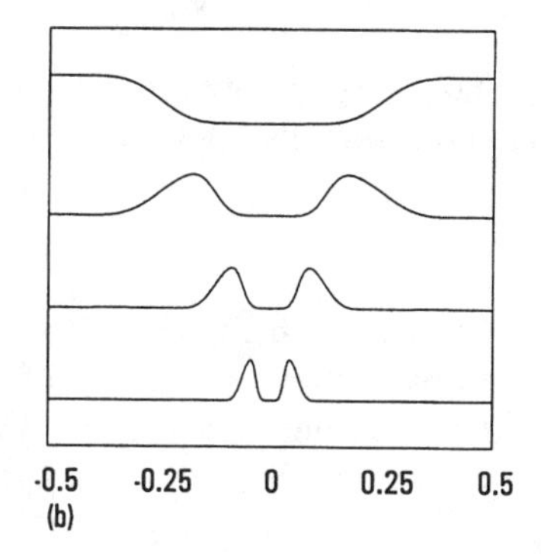

FIGURE 3.4
(a) Spectra of Coif12 scaling signals, from $k = 1$ at top to $k = 4$ at bottom. (b) Spectra of Coif12 wavelets from $k = 1$ at top to $k = 4$ at bottom. The spectra have been multiplied by constants 2^{-k} for each k in order to make their graphs of similar size.

the spectrum of the 1-level wavelet $\mathbf{W}_1^1$.

As with the first level, the DFTs of higher level averaged signals and fluctuation signals can be obtained by multiplying the signal's DFT by the spectra of higher level scaling signals and wavelets. For example, the DFT of the second averaged signal $\mathbf{A}^2$ satisfies

$$\mathcal{F}\mathbf{A}^2 \approx \frac{1}{4} \left|\mathcal{F}\mathbf{V}_1^2\right|^2 \mathcal{F}\mathbf{f} \tag{3.22}$$

and the DFT of the second detail signal $\mathbf{D}^2$ satisfies

$$\mathcal{F}\mathbf{D}^2 \approx \frac{1}{4} \left|\mathcal{F}\mathbf{W}_1^2\right|^2 \mathcal{F}\mathbf{f}. \tag{3.23}$$

In Formula (3.22), 1/4 times the spectrum $\left|\mathcal{F}\mathbf{V}_1^2\right|^2$ of the 2-level scaling signal $\mathbf{V}_1^2$ acts as a low-pass filter. This low-pass filter is graphed in Figure 3.4(a) as the second graph from the top. Notice that this low-pass filter allows through only much lower frequencies than the first-level filtering due to $\left|\mathcal{F}\mathbf{V}_1^1\right|^2$. The spectrum $\left|\mathcal{F}\mathbf{W}_1^2\right|^2$ in Formula (3.23), however, does not act as a high-pass filter. As can be seen from the second graph at the top of Figure 3.4(b), 1/4 times $\left|\mathcal{F}\mathbf{W}_1^2\right|^2$ acts as a *band-pass* filter, in the sense that only two small isolated bands of frequency values are non-zero for this spectrum.

Higher level MRAs follow this same pattern. The k-level averaged signal $\mathbf{A}^k$ has a DFT which satisfies

$$\mathcal{F}\mathbf{A}^k \approx \frac{1}{2^k} \left|\mathcal{F}\mathbf{V}_1^k\right|^2 \mathcal{F}\mathbf{f}. \tag{3.24}$$

And the signal $\mathbf{L}_k = 2^{-k} \left|\mathcal{F}\mathbf{V}_1^k\right|^2$ acts as a low-pass filter. In Figure 3.4(a) we show these low-pass filters $\mathbf{L}_k$ for $k = 3$ and $k = 4$. Notice that only a very small interval of values around the origin are non-zero for $\mathbf{L}_4$. The k-level detail signal's DFT satisfies

$$\mathcal{F}\mathbf{D}^k \approx \frac{1}{2^k} \left|\mathcal{F}\mathbf{W}_1^k\right|^2 \mathcal{F}\mathbf{f}. \tag{3.25}$$

As can be seen from Figure 3.4(b), for $k > 1$, the signal $2^{-k} \left|\mathcal{F}\mathbf{W}_1^k\right|^2$ acts as a band-pass filter. The bands of non-zero values for this filter consist of the values lying between the downward sloping left and right sides of the central portions of the graphs of the low-pass filters $\mathbf{L}_k$ and $\mathbf{L}_{k-1}$.

Low-pass and high-pass filtering *

In this optional subsection we shall discuss the mathematical formulation of the low-pass and high-pass filtering interpretation of wavelet MRAs. Our tool will be the z-transform described at the end of Section 3.2.

From Formula (2.13b), we have

$$\mathbf{A}^1 = a_1\mathbf{V}_1^1 + a_2\mathbf{V}_2^1 + \cdots + a_{N/2}\mathbf{V}_{N/2}^1 \tag{3.26}$$

where $(a_1, a_2, \ldots, a_{N/2})$ is the first trend subsignal $\mathbf{a}^1$. Formula (3.14) tells us that to obtain the DFT, $\mathcal{F}\mathbf{A}^1$, of $\mathbf{A}^1$ it suffices to obtain its z-transform $\mathbf{A}^1[z]$. By Formulas (3.26) and (3.18) we have

$$\begin{aligned}\mathbf{A}^1[z] &= a_1\mathbf{V}_1^1[z] + a_2\mathbf{V}_2^1[z] + \cdots + a_{N/2}\mathbf{V}_{N/2}^1[z] \\ &= a_1\mathbf{V}_1^1[z] + a_2\mathbf{V}_1^1[z]z^2 + \cdots + a_{N/2}\mathbf{V}_1^1[z]z^{N-2}.\end{aligned}$$

Thus $\mathbf{A}^1[z]$ satisfies

$$\mathbf{A}^1[z] = \mathbf{V}_1^1[z]\left(a_1 + a_2z^2 + \cdots + a_{N/2}z^{N-2}\right). \tag{3.27}$$

Because of Formula (3.27) we see that the DFT of $\mathbf{A}^1$ is obtained as a product of the DFT of $\mathbf{V}_1^1$ with the DFT obtained from the polynomial

$$a_1 + a_2z^2 + \cdots + a_{N/2}z^{N-2} = \mathbf{a}^1[z^2].$$

It now remains to examine the connection between the polynomial $\mathbf{a}^1[z^2]$ and the z-transform of the signal $\mathbf{f}$.

The polynomial $\mathbf{a}^1[z^2]$ satisfies

$$\begin{aligned}\mathbf{a}^1[z^2] &= a_1 + a_2z^2 + \cdots + a_{N/2}z^{N-2} \\ &= (\mathbf{f}\cdot\mathbf{V}_1^1) + (\mathbf{f}\cdot\mathbf{V}_2^1)z^2 + \cdots + (\mathbf{f}\cdot\mathbf{V}_{N/2}^1)z^{N-2} \\ &= \sum_{m=1}^{N/2} (\mathbf{f}\cdot\mathbf{V}_m^1)z^{2m-2} \\ &= \sum_{m=1}^{N/2} (\mathbf{f}\cdot\mathcal{T}_{2m-2}\mathbf{V}_1^1)z^{2m-2}.\end{aligned} \tag{3.28}$$

This last polynomial in z consists of the even powered terms from the polynomial on the left side of the following identity:

$$\sum_{k=1}^{N}(\mathbf{f}\cdot\mathcal{T}_{k-1}\mathbf{V}_1^1)z^{k-1} = \mathbf{f}[z]\mathbf{V}_1^1[z^{-1}]. \tag{3.29}$$

The fact that (3.29) holds is a consequence of the definition of multiplication of polynomials; we leave its proof as an exercise for the reader.

The sum of even powered terms can be extracted from the polynomial on the right side of (3.29) by the following identity:

$$\mathbf{f}[z]\mathbf{V}_1^1[z^{-1}] + \mathbf{f}[-z]\mathbf{V}_1^1[-z^{-1}] = 2\sum_{m=1}^{N/2}(\mathbf{f}\cdot\mathcal{T}_{2m-2}\mathbf{V}_1^1)z^{2m-2}. \tag{3.30}$$

Combining (3.30) with (3.28) yields

$$\mathbf{a}^1[z^2] = \frac{1}{2}\mathbf{f}[z]\mathbf{V}_1^1[z^{-1}] + \frac{1}{2}\mathbf{f}[-z]\mathbf{V}_1^1[-z^{-1}]. \tag{3.31}$$

Formula (3.31) is the desired relation between the polynomial $\mathbf{a}^1[z^2]$ and the z-transform of $\mathbf{f}$. Combining it with (3.27) yields

$$\mathbf{A}^1[z] = \frac{1}{2}\mathbf{f}[z]\mathbf{V}_1^1[z]\mathbf{V}_1^1[z^{-1}] + \frac{1}{2}\mathbf{f}[-z]\mathbf{V}_1^1[z]\mathbf{V}_1^1[-z^{-1}]. \tag{3.32}$$

In order to interpret Equation (3.32) correctly, we need to understand the effects of substituting z^{-1} and $-z$ into z-transforms. Substituting z^{-1} into the z-transform $\mathbf{f}[z]$ yields

$$\mathbf{f}[z^{-1}] = \sum_{m=1}^{N} f_m(z^{-1})^{m-1}.$$

We shall use the overbar notation, $\overline{}$, to denote the conjugation of complex numbers. Since z is on the unit-circle, we have $z^{-1} = \overline{z}$. And, because each value f_m of $\mathbf{f}$ is a real number, we have $f_m = \overline{f_m}$. Therefore

$$\begin{aligned}\mathbf{f}[z^{-1}] &= \sum_{m=1}^{N} \overline{f_m}\,\overline{z^{m-1}} \\ &= \overline{\mathbf{f}[z]}.\end{aligned}$$

Thus, for example, $\mathbf{V}_1^1[z]\mathbf{V}_1^1[z^{-1}] = \left|\mathbf{V}_1^1[z]\right|^2$.

To interpret the substitution of $-z$, we observe that $-z$ is the reflection of z through the origin of the complex plane. Consequently, the low frequency values that lie near $e^{i0} = 1$ are reflected into high frequency values that lie near $e^{\pm i\pi} = -1$, and vice versa. For example, in Figure 3.2, the spectrum $|\mathbf{V}_1^1[-z]|^2$ has a graph that is similar to the graph of the spectrum $|\mathbf{W}_1^1[z]|^2$ shown in Figure 3.2(b). The graph of $|\mathbf{V}_1^1[-z^{-1}]|^2$ is identical to the graph of $|\mathbf{V}_1^1[-z]|^2$, because the substitution of z^{-1} produces a complex conjugate which is then eliminated by the modulus-square operation. These considerations show that for the Coif12 scaling function, we have

$$|\mathbf{V}_1^1[z]|\,|\mathbf{V}^1[-z^{-1}]| \approx 0 \tag{3.33}$$

except for two small intervals of values of z centered on $e^{\pm i\pi/2}$. The approximation in (3.33) is true for all of the Daubechies scaling functions.

Based on (3.32) and (3.33) we have the following approximation:

$$\mathbf{A}^1[z] \approx \frac{1}{2}\,|\mathbf{V}_1^1[z]|^2\,\mathbf{f}[z]. \tag{3.34}$$

Using the connection between DFTs and z-transforms, the approximation (3.20) follows from the approximation (3.34).

Similar calculations for the first-level detail signal $\mathbf{D}^1$ and fluctuation subsignal $\mathbf{d}^1$ yield

$$\mathbf{D}^1[z] = \frac{1}{2}\mathbf{f}[z]\mathbf{W}_1^1[z]\mathbf{W}_1^1[z^{-1}] + \frac{1}{2}\mathbf{f}[-z]\mathbf{W}_1^1[z]\mathbf{W}_1^1[-z^{-1}] \tag{3.35}$$

and

$$\mathbf{d}^1[z] = \frac{1}{2}\mathbf{f}[z]\mathbf{W}_1^1[z^{-1}] + \frac{1}{2}\mathbf{f}[-z]\mathbf{W}_1^1[-z^{-1}]. \tag{3.36}$$

And we also have the approximation

$$\mathbf{D}^1[z] \approx \frac{1}{2}\,|\mathbf{W}_1^1[z]|^2\,\mathbf{f}[z], \tag{3.37}$$

which implies the approximation (3.21).

The other approximations, (3.22) through (3.25), can be proved in the same way as (3.20) and (3.21).

3.4 Correlation and feature detection

In this section we shall describe a standard method for detecting a short-lived feature within a more complicated signal. This method, known as correlation, is a fundamental part of Fourier analysis. We shall also describe some of the ways in which wavelet analysis can be used to enhance the basic correlation method for feature detection.

Let's begin by examining feature detection for 1D signals. Feature detection is important in seismology, where there is a need to identify characteristic features that indicate, say, earthquake tremors within a long seismological signal. Or, in an electrocardiogram (ECG), it might be necessary to identify portions of the ECG that indicate an abnormal heartbeat.

At the top of Figure 3.5(a) we show a simulated ECG, which we shall refer to as Signal C. The feature that we wish to locate within Signal C is shown in the middle of Figure 3.5(a); this feature is meant to simulate an abnormal heartbeat. It is, of course, easy for us to visually locate the abnormal heartbeat within Signal C, but that is a far cry from an algorithm that a computer could use for automatic detection.

As noted above, the standard method used for feature detection is correlation. The *correlation* of a signal $\mathbf{f}$ with a signal $\mathbf{g}$, both having lengths of N values, will be denoted by $(\mathbf{f}:\mathbf{g})$. It is also a signal of length N, and its kth value $(\mathbf{f}:\mathbf{g})_k$ is defined by

$$(\mathbf{f}:\mathbf{g})_k = f_1 g_k + f_2 g_{k+1} + \cdots + f_N g_{k+N-1}. \tag{3.38}$$

In order for the sum in (3.38) to make sense, the signal $\mathbf{g}$ needs to be periodically extended, i.e., we assume that $g_{k+N} = g_k$ for each k. When computing the correlation $(\mathbf{f}:\mathbf{g})$, the signal $\mathbf{f}$ is the feature that we wish to detect within the signal $\mathbf{g}$. Usually the signal $\mathbf{f}$ is similar to the abnormal heartbeat signal shown in the middle of Figure 3.5(a), in the sense that the values of $\mathbf{f}$ are 0 except near the central portion of the signal. This reduces the distortion that results from assuming that $\mathbf{g}$ is periodic. We will show later how correlations are related to Fourier analysis, but first we shall describe their use in feature detection.

The rationale behind using the correlation $(\mathbf{f}:\mathbf{g})$ to detect the location of $\mathbf{f}$ within $\mathbf{g}$ is the following. If a portion of $\mathbf{g}$ matches the form of the central portion of $\mathbf{f}$—where the significant, non-zero values are concentrated—then, for a certain value of k, the terms in (3.38) will all be squares. This produces a positive sum which is generally larger than the sums for the other values of $(\mathbf{f}:\mathbf{g})$. In order to normalize this largest value so that it equals 1, we shall divide the values of $(\mathbf{f}:\mathbf{g})$ by the energy of $\mathbf{f}$. That is, we denote the *normalized correlation* of $\mathbf{f}$ with $\mathbf{g}$ by $\langle\mathbf{f}:\mathbf{g}\rangle$, and the kth value of $\langle\mathbf{f}:\mathbf{g}\rangle$ is

$$\langle\mathbf{f}:\mathbf{g}\rangle_k = \frac{f_1 g_k + f_2 g_{k+1} + \cdots + f_N g_{k+N-1}}{\mathcal{E}_{\mathbf{f}}}. \tag{3.39}$$

At the end of this section we shall discuss why, under the right conditions, the maximum value for $\langle\mathbf{f}:\mathbf{g}\rangle$ is approximately 1.

As an example of these ideas, we show at the bottom of Figure 3.5(a) the graph of the squares of the positive values of the normalized correlation $\langle\mathbf{f}:\mathbf{g}\rangle$ for the abnormal heartbeat and Signal C. Notice how *the maximum for this graph clearly locates the position of the heartbeat within Signal C.*

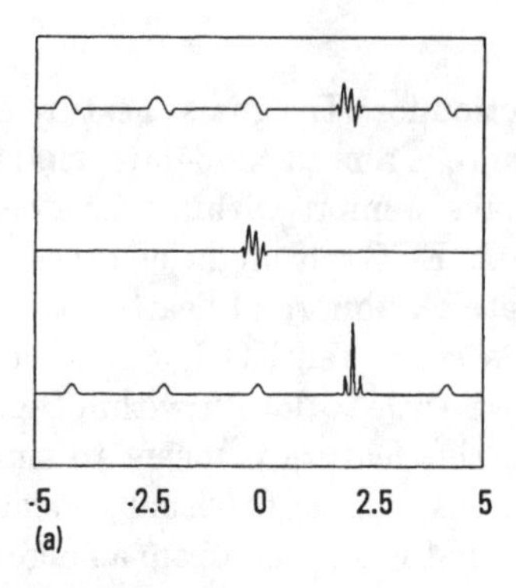

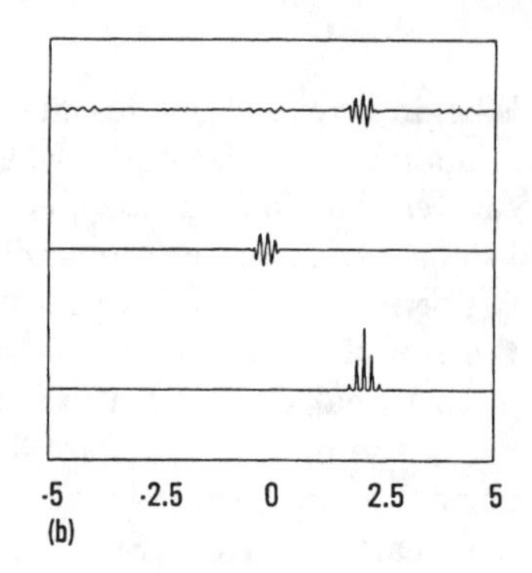

FIGURE 3.5
(a) Top: Signal C. Middle: Abnormal heartbeat. Bottom: Squares of positive values of normalized correlation. (b) Top: Signal C minus its fourth Coif30 averaged signal. Middle: Abnormal heartbeat minus its fourth Coif30 averaged signal. Bottom: Squares of positive values of normalized correlation.

The value of this maximum is 1, thus providing the following simple criterion for locating an abnormal heartbeat: *if a normalized correlation value is near* 1, *then an abnormal heartbeat is probably present at the location of this value.* We have ignored the negative values of the normalized correlation because a negative value of $\langle \mathbf{f} : \mathbf{g} \rangle$ indicates a preponderance of oppositely signed values, which is a clear indication that the values of $\mathbf{f}$ and $\mathbf{g}$ are not matched. The squaring serves to emphasize the maximum value near 1. It is not necessary, but produces a more easily interpretable graph—a graph for which the maximum value near 1 more clearly stands out from smaller values.

Notice that there are smaller peaks in the bottom graph of Figure 3.5(a) that mark the locations of the normal heartbeats in Signal C. These smaller peaks are present because the abnormal heartbeat was created by forming a sum of a normal heartbeat plus a high frequency "noise" term. Consequently, these peaks reflect a partial correspondence between the normal heartbeat term and each of the normal heartbeats in Signal C.

We shall now describe a wavelet based method for suppressing these peaks in the detection signal resulting from the normal heartbeats. While this may not be necessary for the case of Signal C, it might be desired for other signals. Furthermore, the method is a simpler 1D version of the 2D method that we shall describe in the next section.

Each normal heartbeat in Signal C has a spectrum that has significant values only for very low frequencies in comparison to the high frequency oscillations that are clearly visible in the abnormal heartbeat. Our method consists, therefore, of subtracting away an averaged signal from Signal C. As we saw in the previous section, the higher the level k of the averaged signal

$\mathbf{A}^k$, the nearer to zero are the low frequencies which make up the non-zero values of the averaged signal's spectrum $|\mathcal{F}\mathbf{A}^k|^2$. Hence, by subtracting away from Signal C an averaged signal $\mathbf{A}^k$ for high enough k, we can remove the low frequency values from the spectrum of $\mathbf{f}$ that arise from the normal heartbeats.

For example, at the top of Figure 3.5(b) we show the signal that is the difference between Signal C and its fourth averaged signal $\mathbf{A}^4$. Comparing this signal with Signal C we can see that the normal heartbeats have been removed, but there is still a residue corresponding to the abnormal heartbeat. In the middle of Figure 3.5(b) we show the signal equal to the difference between the abnormal heartbeat and its fourth averaged signal, which is a match to the residue of the abnormal heartbeat in the signal at the top of the figure. Finally, at the bottom of Figure 3.5(b) we show the graph of all the squares of the positive values of the normalized correlation of the middle signal with the top signal. This signal clearly locates the position of the abnormal heartbeat, at the same location as before, but without the secondary peaks for the normal heartbeats.

This wavelet method may strike some readers as contrived; but, in the next section we shall use this same method on 2D images of real scenes and achieve the same successful results. We introduced the method here because it is easier to understand its frequency interpretation in the 1D case.

DFT method of computing correlations

There is a simple relationship between correlations and the DFTs of signals. We shall now give a brief sketch of this relationship. Since the normalized correlation $\langle \mathbf{f} : \mathbf{g} \rangle$ simply consists of dividing $(\mathbf{f} : \mathbf{g})$ by the energy of $\mathbf{f}$, we shall concentrate on the problem of computing $(\mathbf{f} : \mathbf{g})$.

The following formula, which we shall prove later, describes the relationship between the DFT of $(\mathbf{f} : \mathbf{g})$ and the DFTs of $\mathbf{f}$ and $\mathbf{g}$:

$$(\mathbf{f} : \mathbf{g}) \overset{\mathcal{F}}{\longmapsto} \mathcal{F}\mathbf{f}\, \overline{\mathcal{F}\mathbf{g}} \tag{3.40}$$

where $\mathcal{F}\mathbf{f}\, \overline{\mathcal{F}\mathbf{g}}$ is the product signal with values $\left(\mathcal{F}\mathbf{f}\, \overline{\mathcal{F}\mathbf{g}}\right)_n = (\mathcal{F}\mathbf{f})_n\, \overline{(\mathcal{F}\mathbf{g})_n}$. This formula shows that the frequency content of $(\mathbf{f} : \mathbf{g})$ simply consists of the product of the DFT of $\mathbf{f}$ with the complex conjugate of the DFT of $\mathbf{g}$. Formula (3.40) also gives us the following three-step method for computing the correlation $(\mathbf{f} : \mathbf{g})$.

DFT calculation of correlation $(\mathbf{f} : \mathbf{g})$

Step 1. Compute DFTs of $\mathbf{f}$ and $\mathbf{g}$.

Step 2. Multiply the values of $\mathcal{F}\mathbf{f}$ and $\overline{\mathcal{F}\mathbf{g}}$ to produce $\mathcal{F}\mathbf{f}\, \overline{\mathcal{F}\mathbf{g}}$.

Step 3. Compute the DFT inverse of $\mathcal{F}\mathbf{f}\,\overline{\mathcal{F}\mathbf{g}}$ to produce $(\mathbf{f}:\mathbf{g})$.

Although this method may appear convoluted at first sight, it is actually much more efficient than a direct calculation of $(\mathbf{f}:\mathbf{g})$ based on Formula (3.38).

The reason that the DFT calculation of correlations is more efficient than Formula (3.38) is because an FFT algorithm is employed for performing the DFTs. Formula (3.38) requires N multiplications and $N-1$ additions in order to calculate each of the N values of $(\mathbf{f}:\mathbf{g})$. That amounts to $2N^2 - N$ operations in total to compute $(\mathbf{f}:\mathbf{g})$. Computing the correlation $(\mathbf{f}:\mathbf{g})$ by this direct method is said to require $O(N^2)$ operations; meaning that a bounded multiple of N^2 operations is needed. An FFT algorithm requires only $O(N \log_2 N)$ operations, hence the DFT calculation of correlation also requires only $O(N \log_2 N)$ operations. When N is large, say $N \geq 1024$, then $N \log_2 N$ is significantly smaller than N^2.

The DFT calculation of correlation is even more efficient when 2D images are involved, as in the next section. For 2D images, say both N by N, the correlation defined in (3.48) in the next section requires $O(N^4)$ operations if a direct calculation is performed. A DFT calculation, however, requires only $O(N^2 \log_2 N)$ operations. The value of N need not be very large at all in order for $N^2 \log_2 N$ to be significantly smaller than N^4.

Proof of (3.40) *

Formula (3.40) can be proved easily using z-transforms. Making use of cyclic translation [see Section 3.2], we can rewrite (3.38) in the form

$$(\mathbf{f}:\mathbf{g})_k = \mathbf{f} \cdot \mathcal{T}_{k-1}\mathbf{g}. \tag{3.41}$$

Consequently, the z-transform of $(\mathbf{f}:\mathbf{g})$ is

$$\begin{aligned}(\mathbf{f}:\mathbf{g})[z] &= \sum_{k=1}^{N} (\mathbf{f} \cdot \mathcal{T}_{k-1}\mathbf{g})\, z^{k-1} \\ &= \mathbf{f}[z]\, \mathbf{g}[z^{-1}].\end{aligned} \tag{3.42}$$

The second equality in (3.42) holds for the same reason that Equality (3.29) holds. Since $\mathbf{g}$ is a real-valued signal, we have $\mathbf{g}[z^{-1}] = \overline{\mathbf{g}[z]}$; so (3.42) becomes

$$(\mathbf{f}:\mathbf{g})[z] = \mathbf{f}[z]\, \overline{\mathbf{g}[z]}. \tag{3.43}$$

From (3.43) and the relation between z-transforms and DFTs, we obtain Formula (3.40).

Normalized correlations and feature detection *

In this subsection we shall briefly examine the mathematical justification for using normalized correlations to detect the presence of one signal within another, more complicated, signal. This discussion will make use of concepts from linear algebra—in particular, Cauchy's inequality for scalar products. Those readers who are not conversant with linear algebra should feel free to skip this subsection; we shall not be referring to it in the sequel.

Let $\mathbf{f}$ be a signal of positive energy, $\mathcal{E}_{\mathbf{f}} > 0$. We assume a positive energy in order to force $\mathbf{f} \neq (0, 0, \ldots, 0)$, because it is clearly pointless to try to detect the signal $(0, 0, \ldots, 0)$ within any signal. Suppose that the signal $\mathbf{g}$ contains the signal $\mathbf{f}$ in the sense that, for some integer m between 1 and N,

$$\mathbf{g} = \mathcal{T}_{m-1}\mathbf{f} + \mathbf{n} \tag{3.44}$$

where $\mathcal{T}_{m-1}\mathbf{f}$ is a cyclic translate of $\mathbf{f}$ and $\mathbf{n}$ is a *noise term*. By noise term we mean an undesired portion of the signal. The signal $\mathbf{n}$ is certainly undesired because we want to detect the presence of $\mathbf{f}$; however, the detection method based on normalized correlation works best when $\mathbf{n}$ is, in fact, random noise that is completely uncorrelated to $\mathbf{f}$. By completely uncorrelated to $\mathbf{f}$ we mean that

$$\langle \mathbf{f} : \mathbf{n} \rangle_j = \frac{\mathbf{f} \cdot \mathcal{T}_{j-1}\mathbf{n}}{\mathcal{E}_{\mathbf{f}}} \approx 0 \tag{3.45}$$

holds for each integer j. Of course, as we saw with the example of Signal C, it is not absolutely necessary that (3.45) holds; but it makes a more general derivation easier, and this derivation provides some insight into the case of Signal C as well.

In any case, assuming that (3.45) holds, we now let $k = N - m + 2$ and note that $\mathcal{T}_{k-1} \circ \mathcal{T}_{m-1} = \mathcal{T}_N = \mathcal{T}_0$. Since $\mathcal{T}_0$ is the identity mapping, we then have

$$\begin{aligned} \langle \mathbf{f} : \mathbf{g} \rangle_k &= \frac{\mathbf{f} \cdot (\mathcal{T}_{k-1} \circ \mathcal{T}_{m-1}\mathbf{f})}{\mathcal{E}_{\mathbf{f}}} + \frac{\mathbf{f} \cdot \mathcal{T}_{k-1}\mathbf{n}}{\mathcal{E}_{\mathbf{f}}} \\ &= \frac{\mathbf{f} \cdot \mathbf{f}}{\mathcal{E}_{\mathbf{f}}} + \frac{\mathbf{f} \cdot \mathcal{T}_{k-1}\mathbf{n}}{\mathcal{E}_{\mathbf{f}}}. \end{aligned}$$

Because $\mathbf{f} \cdot \mathbf{f} = \mathcal{E}_{\mathbf{f}}$ and (3.45) holds, we then have

$$\langle \mathbf{f} : \mathbf{g} \rangle_k = 1 \tag{3.46}$$

except for a small error that (3.45) allows us to ignore.

On the other hand, if n is any of the integers between 1 and N, then letting $j = n + m - 2$ yields

$$\begin{aligned} \langle \mathbf{f} : \mathbf{g} \rangle_n &= \frac{\mathbf{f} \cdot \mathcal{T}_j\mathbf{f}}{\mathcal{E}_{\mathbf{f}}} + \frac{\mathbf{f} \cdot \mathcal{T}_{n-1}\mathbf{n}}{\mathcal{E}_{\mathbf{f}}} \\ &\approx \frac{\mathbf{f} \cdot \mathcal{T}_j\mathbf{f}}{\mathcal{E}_{\mathbf{f}}}. \end{aligned}$$

By the Cauchy inequality, we obtain

$$|\mathbf{f} \cdot \mathcal{T}_j \mathbf{f}| \leq \sqrt{\mathcal{E}_{\mathbf{f}}} \sqrt{\mathcal{E}_{\mathcal{T}_j \mathbf{f}}} = \mathcal{E}_{\mathbf{f}}$$

and equality holds if and only if $\mathcal{T}_j \mathbf{f} = \mathbf{f}$ or $\mathcal{T}_j \mathbf{f} = -\mathbf{f}$. Therefore, except for a small error which (3.45) allows us to ignore, we have

$$-1 \leq \langle \mathbf{f} : \mathbf{g} \rangle_n \leq 1. \tag{3.47}$$

Equality holds on the right side of (3.47) only when $\mathbf{f} = \mathcal{T}_j \mathbf{f}$.

This discussion shows that we can detect the presence of cyclic translates of **f** within the signal **g** by the location of maximum value(s) of 1 among the positive values of $\langle \mathbf{f} : \mathbf{g} \rangle$. The method works best when those translates of **f** which are not equal to **f** produce scalar products that are much smaller than the energy of **f**; this is the case, for instance, with the abnormal heartbeat considered above. Furthermore, although (3.45) does not hold for Signal C, the value of $\mathbf{f} \cdot \mathcal{T}_{k-1} \mathbf{n}$ is 0; hence (3.46) still holds. And, for the other index values j, the values of $(\mathbf{f} \cdot \mathcal{T}_j \mathbf{f})/\mathcal{E}_{\mathbf{f}}$ and $(\mathbf{f} \cdot \mathcal{T}_{n-1} \mathbf{n})/\mathcal{E}_{\mathbf{f}}$ are small enough that they add up to less than 1.

The discussion in this subsection can be extended to 2D images and will apply to the wavelet based method of object detection in images that we shall describe in the next section.

3.5 Object detection in 2D images

In the previous section we described a basic method of feature detection, and a wavelet based enhancement of it, for 1D signals. In this section we shall describe how objects can be detected within 2D images. We shall discuss some examples related to character detection and the more difficult problem of locating small objects within complicated scenes.

Our first example shows how to identify the image of the character, P, shown in Gr 1 of Figure 3.6, within the sequence of three characters, PQW, shown in Gr 2 of Figure 3.6. One method for doing this is to compute a normalized correlation. For 2D images **f** and **g**, a normalized correlation of **f** with **g** is defined as a correlation divided by the energy of **f**, as we did for 1D signals in the previous section. To understand the 2D definition of correlation, we rewrite the 1D definition in (3.38) in the following form:

$$(\mathbf{f} : \mathbf{g})_k = \sum_{n=1}^{N} f_n g_{n+k-1}.$$

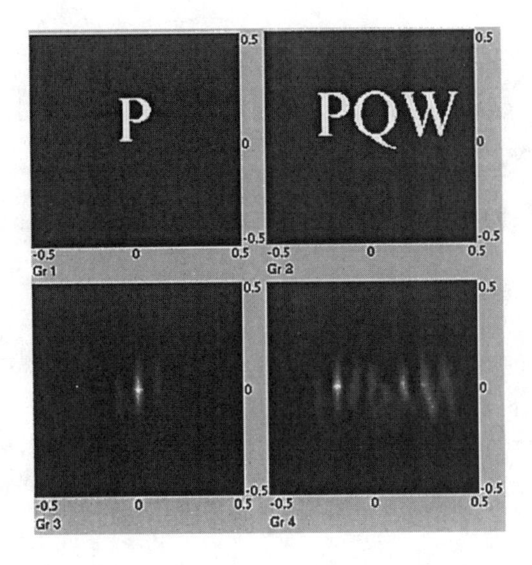

FIGURE 3.6
Gr 1: Image of P. Gr 2: Image of PQW. Gr 3: Squares of positive values of normalized correlation of Gr 1 with itself. Gr 4: Squares of positive values of normalized correlation of Gr 1 with Gr 2.

The 2D definition of the correlation $(\mathbf{f}:\mathbf{g})$ of two M by N images $\mathbf{f}$ and $\mathbf{g}$ is defined by

$$(\mathbf{f}:\mathbf{g})_{k,j} = \sum_{n=1}^{N}\sum_{m=1}^{M} f_{n,m}g_{n+k-1,m+j-1}. \tag{3.48}$$

Formula (3.48) requires a periodic extension of $\mathbf{g}$, i.e., $g_{n+N,m+M} = g_{n,m}$ is assumed to hold for all integers m and n. Based on (3.48), we define the normalized correlation $\langle\mathbf{f}:\mathbf{g}\rangle$ of two M by N images $\mathbf{f}$ and $\mathbf{g}$ by

$$\langle\mathbf{f}:\mathbf{g}\rangle_{k,j} = \frac{(\mathbf{f}:\mathbf{g})_{k,j}}{\mathcal{E}_{\mathbf{f}}}. \tag{3.49}$$

As with 1D signals, the identification of an object $\mathbf{f}$ within an image $\mathbf{g}$ is achieved by locating maximum values that are approximately 1 from among the positive values of $\langle\mathbf{f}:\mathbf{g}\rangle$. For example, if the image of P in Gr 1 of Figure 3.6 is the object $\mathbf{f}$ and the image of PQW in Gr 2 is the image $\mathbf{g}$, then we show the squares of the positive values of $\langle\mathbf{f}:\mathbf{g}\rangle$ in Gr 4. The maximum value of 1 is clearly visible and locates the position of P within the image of PQW.

Our second example involves locating a small object within a complicated scene. This problem reveals the limitations of the correlation method; nevertheless, with some assistance from wavelet analysis, we shall outline an

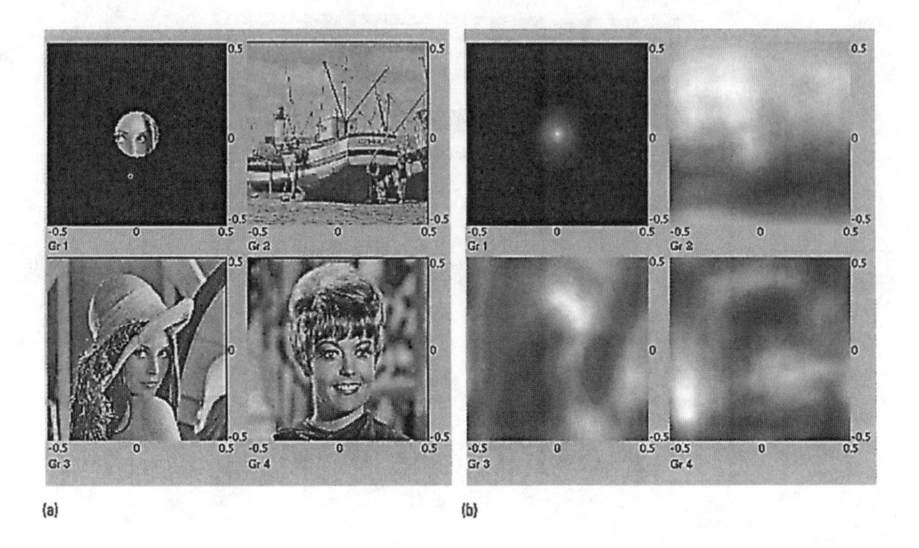

FIGURE 3.7
(a) Four images, Gr 1 is the object that we wish to locate within the other three images. (b) Squares of positive values of normalized correlations of Gr 1 from (a) with each of the images in (a).

effective solution. In Gr 1 of Figure 3.7(a), we show a small image. We wish to identify the location, or the absence, of this object within the other three images in Gr 2 through Gr 4 of Figure 3.7(a). In Figure 3.7(b) we show the squares of the positive values of the normalized correlations of the desired object with each of the images in Figure 3.7(a). It is clear from these Figures that the method fails; it successfully locates Gr 1 within Gr 1, but does not correctly locate Gr 1 within Gr 3. Furthermore, for Gr 2 and Gr 3, the maximum values of the normalized correlations are 1.14 and 1.13. The fact that these maximum values are slightly greater than 1 indicates a partial breakdown of the method, but the worst result is the fact that Gr 1 is not present at all within Gr 2.

We shall now describe a wavelet based approach to identifying the location of this object. This approach rests on the fact that objects can often be identified via their edges; for example, there is some evidence that our visual system works on this basis. We saw at the end of the last chapter that the first-level detail signal $\mathbf{D}^1$ can provide us with an image consisting of edges. Based on these observations, we shall use the following three-step method for locating the object.

Edge Correlation Method of Object Location

Step 1. Compute a first-level detail image $\mathbf{D}^1$ for the object.

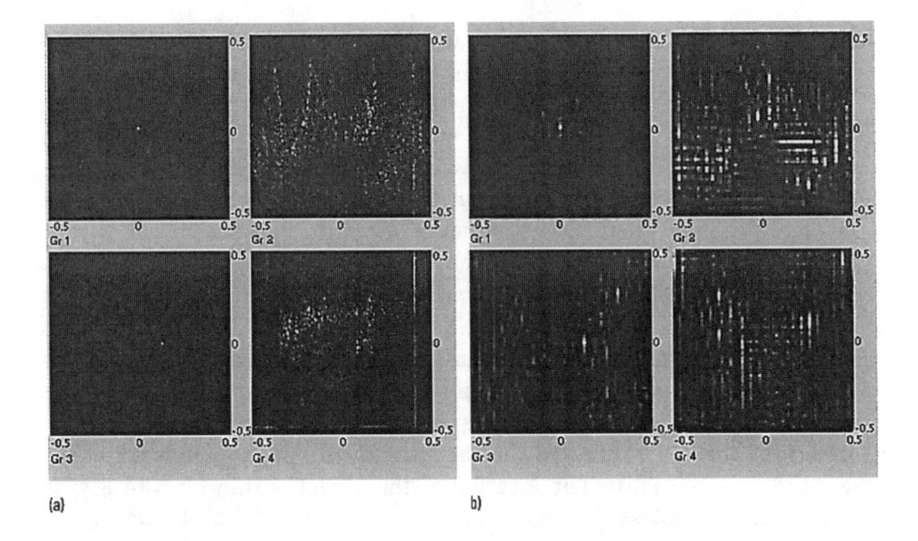

FIGURE 3.8
(a) Squares of positive values of the normalized correlations produced by the Edge Correlation Method applied to the images in Figure 3.7(a). (b) Similarly produced images resulting from the third trends of the images in Figure 3.7(a).

> **Step 2.** Eliminate from this detail image any extraneous edge effects resulting from the outer edges in the object. This is not always necessary, but is needed with the object in Gr 1 in Figure 3.7(a). Removing its outer edge effects can be done by only retaining values that lie within a small enough radius.
>
> **Step 3.** Compute normalized correlations of the image from Step 2 with first-level detail signals for each of the other images. Determine if there are any values near 1 and where they are located.

In Figure 3.8(a) we show the results of this Edge Correlation Method, using Coif12 wavelets to produce the detail images. We have graphed the squares of the positive values of the normalized correlations, which emphasizes a maximum value of 1. The location of the object within Gr 3 in Figure 3.7(a) is clearly indicated. The value of the normalized correlation at this point is 1. For Gr 2 and Gr 4 in Figure 3.8(a), there are no clear indications that the object is located within the corresponding images in Figure 3.7(a). In fact, the maximum values of the normalized correlations are 0.267 and 0.179, respectively, which are significantly less than 1.

We close this section with another example. In the last example, the object was in the same orientation within the image where we located it.

Clearly this would not often be the case. For example, the object might be rotated through some angle, or it might be reflected about some line through the origin, or some combination of these last two operations. Let's refer to these operations as *symmetry operations* on the object. One solution to the problem of locating the object when it is in a different orientation is to perform the Edge Correlation Method on a finite set of objects obtained from performing a large number of symmetry operations on the initial object. The problem with this approach is that it is prohibitively time consuming.

We shall now outline a less time consuming solution to the problem just described. The gist of this solution is to perform the Edge Correlation Method on trend subimages for some specific level of wavelet transforms of the images. For example, in Figure 3.8(b), we show the squares of the positive values of the normalized correlation images produced by the Edge Correlation Method applied to the third trend subimages $\mathbf{a}^3$ of each of the images in Figure 3.7(a). The location of the object within Gr 3 in Figure 3.7(a) is clearly indicated. The value of the normalized correlation at this point is 0.852, which is not 1 but is close to it. For Gr 2 and Gr 4 in Figure 3.8(b), there are no clear indications that the object is located within the corresponding images in Figure 3.7(a). In fact, the maximum values of the normalized correlations are 0.431 and 0.402, respectively, which are significantly less than 1.

The advantage of the method just described is that, by working with third level trend subimages, the sizes of the images are reduced by a factor of 64. This makes it feasible to perform a large number of symmetry operations on the object in order to search for rotated, or reflected, versions of it.

In illustrating the Edge Correlation Method, we ignored one important point. In order for the method to work effectively, the location of the object within an image must be centered on even index values. This is because when the first level subimages are computed, the scaling signals and wavelets are all shifts of $\mathbf{V}_1^1$ and $\mathbf{W}_1^1$ by even integers. Therefore, in general, it is necessary to perform the Edge Correlation Method on three other objects obtained by shifting the initial object by one index value in the horizontal direction, by one index value in the vertical direction, and by one index value in both directions.

3.6 Creating scaling signals and wavelets *

We conclude this chapter with an outline of the way in which scaling signals and wavelets are created from properties of their z-transforms. This material is more difficult than the other material in this chapter, and makes use of some earlier optional material; so those readers who wish to skip over it may certainly do so. No use will be made subsequently of this material.

The heart of wavelet theory is MRA, and so we begin by expressing the 1-level MRA equation $\mathbf{f} = \mathbf{A}^1 + \mathbf{D}^1$ in terms of z-transforms:

$$\mathbf{f}[z] = \mathbf{A}^1[z] + \mathbf{D}^1[z]. \tag{3.50}$$

Using Formulas (3.32) and (3.35) in the right side of (3.50), we obtain

$$\begin{aligned}\mathbf{f}[z] = \mathbf{f}[z]&\left\{\frac{1}{2}\mathbf{V}_1^1[z]\mathbf{V}_1^1[z^{-1}] + \frac{1}{2}\mathbf{W}_1^1[z]\mathbf{W}_1^1[z^{-1}]\right\}\\ &+ \mathbf{f}[-z]\left\{\frac{1}{2}\mathbf{V}_1^1[z]\mathbf{V}_1^1[-z^{-1}] + \frac{1}{2}\mathbf{W}_1^1[z]\mathbf{W}_1^1[-z^{-1}]\right\}. \end{aligned}\tag{3.51}$$

By comparing the two sides of (3.51) we see that the following two equations must hold:

$$\mathbf{V}_1^1[z]\mathbf{V}_1^1[z^{-1}] + \mathbf{W}_1^1[z]\mathbf{W}_1^1[z^{-1}] = 2 \tag{3.52a}$$

$$\mathbf{V}_1^1[z]\mathbf{V}_1^1[-z^{-1}] + \mathbf{W}_1^1[z]\mathbf{W}_1^1[-z^{-1}] = 0. \tag{3.52b}$$

In order to make (3.52b) hold, we define $\mathbf{W}_1^1[z]$ by

$$\mathbf{W}_1^1[z] = -z^{2k+1}\mathbf{V}_1^1[-z^{-1}] \tag{3.53}$$

where the exponent $2k+1$ is an odd integer that we shall specify later.

Before we go further, it is interesting to observe that (3.53) implies that

$$\left|\mathbf{W}_1^1[z]\right| = \left|\mathbf{V}_1^1[-z^{-1}]\right| \tag{3.54}$$

because z is on the unit-circle (so $|z| = 1$). Formula (3.54) implies the approximation (3.33) for the Coif12 case that we considered in Section 3.3, given the graphs of $\left|\mathbf{V}_1^1\right|^2$ and $\left|\mathbf{W}_1^1\right|^2$ shown in Figure 3.2(a).

We now return to our derivation of the z-transforms of scaling functions and wavelets. Combining (3.53) and (3.52a), and the identity

$$\mathbf{V}_1^1[z]\mathbf{V}_1^1[z^{-1}] = \left|\mathbf{V}_1^1[z]\right|^2,$$

we conclude that

$$\left|\mathbf{V}_1^1[z]\right|^2 + \left|\mathbf{V}_1^1[-z]\right|^2 = 2. \tag{3.55}$$

In order to satisfy (3.55) it is easier to work with the function $P(\theta)$ defined by

$$P(\theta) = \frac{1}{\sqrt{2}}\mathbf{V}_1^1[e^{i2\pi\theta}], \tag{3.56}$$

where θ is a real variable. Using this function $P(\theta)$, Equation (3.55) becomes

$$|P(\theta)|^2 + |P(\theta + 1/2)|^2 = 1. \tag{3.57}$$

We will now show how the Daub4 scaling numbers in Equation (2.3) can be obtained by solving Equation (3.57). We begin by observing that the following trigonometric identity

$$|\cos \pi\theta|^2 + |\cos \pi(\theta + 1/2)|^2 = 1 \tag{3.58}$$

resembles the form of (3.57). In fact, if we were to set $P(\theta) = e^{i\pi\theta} \cos \pi\theta$, then we would be led to the two scaling numbers $\alpha_1 = \alpha_2 = 1/\sqrt{2}$ for the Haar scaling function $\mathbf{V}_1^1$. We leave the details to the reader; the reasoning involved is a simplified version of the argument that we shall now use to obtain the Daub4 scaling numbers.

The first step is to cube both sides of (3.58) obtaining

$$\begin{aligned} 1 &= \left(\cos^2 \pi\theta + \sin^2 \pi\theta\right)^3 \\ &= \cos^6 \pi\theta + 3\cos^4 \pi\theta \sin^2 \pi\theta + 3\cos^2 \pi\theta \sin^4 \pi\theta + \sin^6 \pi\theta. \end{aligned} \tag{3.59}$$

We now require that $|P(\theta)|^2$ satisfies

$$|P(\theta)|^2 = \cos^6 \pi\theta + 3\cos^4 \pi\theta \sin^2 \pi\theta, \tag{3.60}$$

which are the first two terms on the right side of (3.59). The remaining two terms on the right side of (3.59) are equal to $|P(\theta + 1/2)|^2$; so (3.57) holds.

Our final task is to obtain a function $P(\theta)$ which satisfies (3.60). In general, this is done via a result known as the *Riesz lemma.* This approach is described in the references on wavelets given at the end of the chapter. For the case of the Daub4 scaling functions, however, we can find $P(\theta)$ by a more direct, though still somewhat tricky, argument. We observe that

$$|P(\theta)|^2 = \cos^4 \pi\theta \left[\cos^2 \pi\theta + 3\sin^2 \pi\theta\right]; \tag{3.61}$$

so we could set $P(\theta) = [\cos \pi\theta]^2 \left[\cos \pi\theta - i\sqrt{3}\sin \pi\theta\right]$. For reasons that will be clear at the end, however, we instead define $P(\theta)$ by

$$P(\theta) = e^{i3\pi\theta} [\cos \pi\theta]^2 \left[\cos \pi\theta - i\sqrt{3}\sin \pi\theta\right]. \tag{3.62}$$

Notice that this formula for $P(\theta)$ implies that $P(\theta)$ satisfies (3.61), because $|e^{i3\pi\theta}|^2 = 1$. Since (3.61) holds, it follows that (3.60) does as well.

Our final task is to convert (3.62) into a form that allows us to read off the z-transform of $\mathbf{V}_1^1[z]$. To do this, we observe that

$$\begin{aligned} P(\theta) &= e^{i6\pi\theta} \left[e^{-i\pi\theta} \cos \pi\theta\right]^2 \left[e^{-i\pi\theta} \cos \pi\theta - i\sqrt{3}e^{-i\pi\theta} \sin \pi\theta\right] \\ &= e^{i6\pi\theta} \frac{1}{4} \left[1 + e^{-i2\pi\theta}\right]^2 \left[\frac{1 + e^{-i2\pi\theta}}{2} + \frac{\sqrt{3}}{2}\left(e^{-i2\pi\theta} - 1\right)\right]. \end{aligned}$$

Multiplying out the last expression and simplifying yields the formula we need:

$$P(\theta) = \frac{1+\sqrt{3}}{8} + \frac{3+\sqrt{3}}{8}\,e^{i2\pi\theta} + \frac{3-\sqrt{3}}{8}\,e^{i4\pi\theta} + \frac{1-\sqrt{3}}{8}\,e^{i6\pi\theta}. \quad (3.63)$$

Since $\mathbf{V}_1^1[e^{i2\pi\theta}] = \sqrt{2}\,P(\theta)$, we obtain the z-transform $\mathbf{V}_1^1[z]$ from Formula (3.63) by setting $z = e^{i2\pi\theta}$:

$$\mathbf{V}_1^1[z] = \frac{1+\sqrt{3}}{4\sqrt{2}} + \frac{3+\sqrt{3}}{4\sqrt{2}}\,z + \frac{3-\sqrt{3}}{4\sqrt{2}}\,z^2 + \frac{1-\sqrt{3}}{4\sqrt{2}}\,z^3. \quad (3.64)$$

Formula (3.64) tells us that $\mathbf{V}_1^1$ is defined by

$$\mathbf{V}_1^1 = (\alpha_1, \alpha_2, \alpha_3, \alpha_4, 0, 0, \ldots, 0)$$

where $\alpha_1, \alpha_2, \alpha_3, \alpha_4$ are the Daub4 scaling numbers defined in Formula (2.3). We have thus shown how those scaling numbers can be obtained via z-transform theory.

The definition of the Daub4 wavelet numbers now follows easily. If we set $k = 1$ in Formula (3.53), then $\mathbf{W}_1^1[z] = -z^3\mathbf{V}_1^1[-z^{-1}]$. This equation combined with (3.64) yields

$$\mathbf{W}_1^1[z] = \frac{1-\sqrt{3}}{4\sqrt{2}} + \frac{\sqrt{3}-3}{4\sqrt{2}}\,z + \frac{3+\sqrt{3}}{4\sqrt{2}}\,z^2 + \frac{-1-\sqrt{3}}{4\sqrt{2}}\,z^3. \quad (3.65)$$

Formula (3.65) implies that

$$\mathbf{W}_1^1 = (\beta_1, \beta_2, \beta_3, \beta_4, 0, 0, \ldots, 0)$$

where $\beta_1, \beta_2, \beta_3, \beta_4$ are the Daub4 wavelet numbers defined in Formula (2.8).

We end this section by noting that other properties of wavelets can be obtained by requiring that the function $P(\theta)$ satisfies certain identities. For instance, the condition that the Daub4 scaling numbers satisfy

$$\alpha_1 + \alpha_2 + \alpha_3 + \alpha_4 = \sqrt{2}$$

is equivalent to the requirement that

$$P(0) = 1. \quad (3.66)$$

Notice that the function $P(\theta)$ defined above does satisfy this requirement. Furthermore, the conditions on the Daub4 wavelet numbers, stated in Equations (2.11) and (2.12), can be easily seen to be equivalent to the two equations

$$\mathbf{W}_1^1[1] = 0, \quad \frac{d}{dz}\mathbf{W}_1^1[1] = 0. \quad (3.67)$$

Tracing back through the definitions, it is not hard to show that these last two equations are equivalent to

$$P(1/2) = 0, \quad \frac{dP}{d\theta}(1/2) = 0. \tag{3.68}$$

These equations are, indeed, satisfied by the function $P(\theta)$ defined above.

Equations (3.66) and (3.68) are important because they show how crucial identities involving the scaling numbers and wavelets can be expressed in terms of values of $P(\theta)$ and its derivatives at $\theta = 0$ and $\theta = 1/2$. For example, the function $P(\theta)$ for the Coif6 case is required to satisfy

$$P(0) = 1, \quad \frac{dP}{d\theta}(0) = 0, \quad \frac{d^2P}{d\theta^2}(0) = 0 \tag{3.69}$$

and

$$P(1/2) = 0, \quad \frac{dP}{d\theta}(1/2) = 0. \tag{3.70}$$

The equations in (3.69) correspond to Equations (2.33a) through (2.33c), while the equations in (3.70) correspond to Equations (2.32a) and (2.32b).

3.7 Notes and references

The DFT and the FFT are described in [BRH], [BRI], [WA1], and [WA2]. Using the DFT for noise removal is described in [WA2] and [WA3], and the use of related transforms for compression is discussed in [WAN].

The detection of abnormal heart signals in ECGs is considerably more complicated than our discussion indicates. Further details on wavelet based methods can be found in the papers [STC] and [AKA].

The construction of the Daub4 scaling numbers and wavelet numbers described in Section 3.6 is adapted from a discussion in [STR], where the Daub6 case is examined. The construction of scaling numbers and wavelet numbers in general, based on the Riesz lemma, is described in [DAU] and [VEK]. Newer methods are described in [SW1] and [SW2]. There is also an excellent discussion of the construction of the Coif$I$ scaling numbers and wavelet numbers in [BGG].

A rather complete characterization of wavelets from the standpoint of their frequency content is given in [HEW].

Chapter 4

Beyond wavelets

> When you have only one way of expressing yourself, you have limits that you don't appreciate. When you get a new way to express yourself, it teaches you that there could be a third or a fourth way. It opens up your eyes to a much broader universe.
>
> *David Donoho*[1]

In this chapter we shall explore some additional topics that extend the basic ideas of wavelet analysis introduced previously. We first describe the theory of *wavelet packet transforms,* which sometimes provide superior performance beyond that provided by wavelet transforms. Then we discuss *continuous wavelet transforms* which are particularly useful for tackling problems in signal recognition, and for performing finely detailed examinations of the structure of signals.

4.1 Wavelet packet transforms

A *wavelet packet transform* is a simple generalization of a wavelet transform. In this section we briefly discuss the definition of wavelet packet transforms, and in the next section examine some examples illustrating their applications.

All wavelet packet transforms are calculated in a similar way. Therefore we shall concentrate initially on the Haar wavelet packet transform, which is the easiest to describe. The Haar wavelet packet transform is usually referred to as the *Walsh transform.* A Walsh transform is calculated by

[1] Donoho's quote is from [BUR].

performing a 1-level Haar transform *on all subsignals,* both trends *and* fluctuations.

For example, consider the signal **f** defined by

$$\mathbf{f} = (2, 4, 6, 8, 10, 12, 14, 16). \tag{4.1}$$

A 1-level Haar transform and a 1-level Walsh transform of **f** are identical, producing the following signal:

$$(3\sqrt{2}, 7\sqrt{2}, 11\sqrt{2}, 15\sqrt{2} \,|\, -\sqrt{2}, -\sqrt{2}, -\sqrt{2}, -\sqrt{2}). \tag{4.2}$$

A 2-level Walsh transform is calculated by performing 1-level Haar transforms on both the trend and the fluctuation subsignals, as follows:

$$\begin{aligned} (3\sqrt{2}, 7\sqrt{2}, 11\sqrt{2}, 15\sqrt{2}) &\overset{\mathbf{H}_1}{\longmapsto} (10, 26 \,|\, -4, -4) \\ (-\sqrt{2}, -\sqrt{2}, -\sqrt{2}, -\sqrt{2}) &\overset{\mathbf{H}_1}{\longmapsto} (-2, -2 \,|\, 0, 0). \end{aligned}$$

Hence the 2-level Walsh transform of the signal **f** is the following signal:

$$(10, 26 \,|\, -4, -4 \,|\, -2, -2 \,|\, 0, 0). \qquad \text{[2-level Walsh]} \tag{4.3}$$

It is interesting to compare this 2-level Walsh transform with the 2-level Haar transform of the signal **f**. The 2-level Haar transform of **f** is the following signal:

$$(10, 26 \,|\, -4, -4 \,|\, -\sqrt{2}, -\sqrt{2}, -\sqrt{2}, -\sqrt{2}). \qquad \text{[2-level Haar]} \tag{4.4}$$

Comparing this Haar transform with the Walsh transform in (4.3), we see that the Walsh transform is slightly more compressed in terms of energy, since the last two values of the Walsh transform are zeros. We could, for example, achieve 25% compression of signal **f** by discarding the two zeros from its 2-level Walsh transform, but we could not discard any zeros from its 2-level Haar transform. Another advantage of the 2-level Walsh transform is that it is more likely that *all* of its non-zero values would stand out from a random noise background, because these values have larger magnitudes than the values of the 2-level Haar transform.

A 3-level Walsh transform is performed by calculating 1-level Haar transforms on each of the four subsignals that make up the 2-level Walsh transform. For example, applying 1-level Haar transforms to each of the four subsignals of the 2-level Walsh transform in (4.3), we obtain

$$\begin{aligned} (10, 26) &\overset{\mathbf{H}_1}{\longmapsto} (18\sqrt{2} \,|\, -8\sqrt{2}), \\ (-4, -4) &\overset{\mathbf{H}_1}{\longmapsto} (-4\sqrt{2} \,|\, 0), \\ (-2, -2) &\overset{\mathbf{H}_1}{\longmapsto} (-2\sqrt{2} \,|\, 0), \\ (0, 0) &\overset{\mathbf{H}_1}{\longmapsto} (0 \,|\, 0). \end{aligned}$$

Hence the 3-level Walsh transform of the signal $\mathbf{f}$ in (4.1) is

$$(18\sqrt{2}\,|\,-8\sqrt{2}\,|\,-4\sqrt{2}\,|\,0\,|\,-2\sqrt{2}\,|\,0\,|\,0\,|\,0). \qquad \text{[3-level Walsh]} \quad (4.5)$$

Here, at the third level, the contrast between the Haar and Walsh transforms is even sharper than at the second level. The 3-level Haar transform of this signal is

$$(18\sqrt{2}\,|\,-8\sqrt{2}\,|\,-4, -4\,|\,-\sqrt{2}, -\sqrt{2}, -\sqrt{2}, -\sqrt{2}). \qquad \text{[3-level Haar]} \quad (4.6)$$

Comparing the transforms in (4.5) and (4.6) we can see, at least for this particular signal $\mathbf{f}$, that the 3-level Walsh transform achieves a more compact redistribution of the energy of the signal than the Haar transform.

In general, a wavelet packet transform is performed by calculating a particular 1-level wavelet transform for each of the subsignals of the preceding level. For instance, a 3-level Daub4 wavelet packet transform would be calculated in the same way as a 3-level Walsh transform, but with 1-level Daub4 wavelet transforms being used instead of 1-level Haar transforms. Because all of the 1-level wavelet transforms that we have discussed enjoy the Conservation of Energy property and have inverses, it follows that all of their wavelet packet transforms also enjoy the Conservation of Energy property and have inverses. What this implies is that our discussions of compression and denoising of signals in Chapters 1 and 2 apply, essentially without change, to wavelet packet transforms. In particular, the threshold method is still the basic method for compression and noise removal with wavelet packet transforms.

In two dimensions, a wavelet packet transform is performed by adopting the same approach that we used in one dimension. First, a 1-level wavelet transform is performed on the 2D image. Then, to compute a 2-level wavelet packet transform, this 1-level wavelet transform is applied to each of the four subimages—$\mathbf{a}^1$, $\mathbf{d}^1$, $\mathbf{h}^1$, and $\mathbf{v}^1$—from the first level. This produces 16 subimages that constitute the 2-level wavelet packet transform of the image. To compute a 3-level wavelet packet transform, the 1-level wavelet transform is applied to each of these 16 subimages, producing 64 subimages. This process continues in an obvious manner for higher level wavelet packet transforms.

Because of the great similarity between wavelet transforms and wavelet packet transforms, we shall now end our discussion of the mathematics of these transforms, and turn to a discussion of a few examples of how they can be applied.

4.2 Applications of wavelet packet transforms

In this section we shall discuss a few examples of applying wavelet packet transforms to audio and image compression. While wavelet packet transforms can be used for other purposes, such as noise removal, because of space limitations we shall limit our discussion to the arena of compression.

For our first example, we shall use a Coif30 wavelet packet transform to compress the audio signal *greasy,* considered previously in Section 2.5. In that section we found that a 4-level Coif30 wavelet transform—with trend values quantized at 8 bpp and fluctuations quantized at 6 bpp, and with separate entropies computed for all subsignals—achieved a compression of *greasy* requiring an estimated 11,305 bits. That is, this compression required an estimated 0.69 bpp (instead of 8 bpp in the original). However, if we use a 4-level Coif18 wavelet packet transform and quantize in the same way, then the estimated number of bits is 10,158, i.e., 0.62 bpp. This represents a slight improvement over the wavelet transform.

In several respects—in bpp, in RMS Error, and in total number of significant values—the wavelet packet compression of *greasy* is nearly as good as or slightly better than the wavelet transform compression. See Table 4.1. Whether these slight improvements are worth the extra computations needed to calculate with the wavelet packet transform is certainly open to question. Our next example, from the field of image compression, is more definitive.

Table 4.1 Wavelet and wavelet packet compressions of *greasy*

Transform	*Sig. values*	*Bpp*	*RMS Error*
wavelet	3685	0.69	0.839
w. packet	3072	0.62	0.868

Our second example deals with an image compression. In Figure 4.1(a) we show an image of a woman, which we shall refer to as *Barb.* If a 4-level Coif12 wavelet transform is applied to this image—with the trend quantized at 8 bpp, the fluctuations quantized at 6 bpp, and separate entropies computed for each subimage—then an estimated 0.67 bpp are needed to encode the compressed image. The compressed image is virtually indistinguishable from the original image; so we will not display it. There are some noticeable differences in details at sufficiently high magnification, as can be seen by comparing Figures 4.1(b) and 4.1(c).

If we compute a 4-level Coif12 wavelet packet transform, using the same quantizations as for the wavelet transform, then the compressed image re-

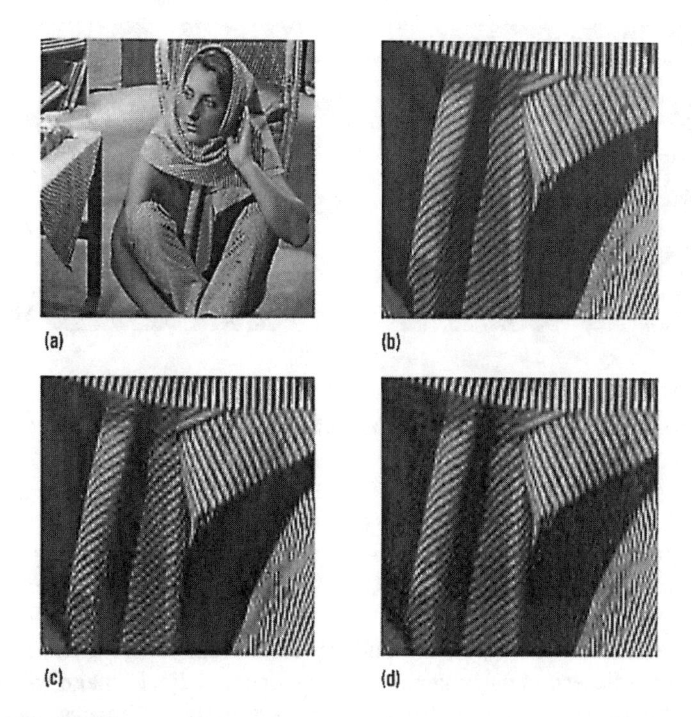

FIGURE 4.1
(a) *Barb* image. (b) Detail of original. (c) Detail of wavelet compressed image. (d) Detail of wavelet packet compressed image.

quires an estimated 0.51 bpp for encoding. This represents a 24% improvement over the wavelet transform compression. The wavelet packet transform performs significantly better in several respects, as summarized in Table 4.2.

Table 4.2 Wavelet and wavelet packet compressions of *Barb*

Transform	*Sig. values*	*Bpp*	*Rel. 2-norm error*
wavelet	28370	0.67	.0486
w. packet	21755	0.51	.0462

There is also improved accuracy of detail in the wavelet packet compression, as shown in Figure 4.1(d). In particular, the two sets of diagonal stripes aligned along the two vertical folds of *Barb*'s scarf are better preserved in Figure 4.1(d) than in Figure 4.1(c).

There is insufficient space to pursue a thorough explanation of why the wavelet packet transform performs better in this example. Nevertheless,

FIGURE 4.2
(a) 2-level Coif12 wavelet transform of *Barb*. (b) 2-level Coif12 wavelet packet transform of *Barb*.

we can gain some understanding of the situation by comparing a Coif12 wavelet transform of *Barb* with a Coif12 wavelet packet transform. In Figure 4.2(a) we show a 2-level Coif12 wavelet transform of *Barb*. Notice that the 1-level fluctuations—$\mathbf{h}^1$, $\mathbf{d}^1$, and $\mathbf{v}^1$—all reveal considerable structure. It is easy to discern ghostly versions of the original *Barb* image within each of these fluctuations. The presence of these ghostly subimages, along with the trend subimage, suggests the possibility of performing wavelet transform compression on all of these subimages. In other words, we compute another 1-level wavelet transform on each of the four subimages, which produces the 2-level wavelet packet transform shown in Figure 4.2(b). Notice that, in the regions corresponding to the horizontal fluctuation $\mathbf{h}^1$ and the diagonal fluctuation $\mathbf{d}^1$ in the wavelet transform, there is a considerable reduction in the number of significant values in the wavelet packet transform. This reduction enables a greater compression of the *Barb* image. For similar reasons, the 4-level wavelet packet transform exhibits a more compact distribution of significant coefficients, hence a greater compression, than the 4-level wavelet transform.

For our last example, we consider a compression of a fingerprint image. In the previous example, we saw that a 4-level wavelet packet transform performed better on the *Barb* image than a 4-level wavelet transform. Consequently, we are led to try a similar compression of Fingerprint 1 [see Figure 2.18(a)]. Instead of the 4-level Coif18 wavelet transform used on Fingerprint 1 in Section 2.9, here we shall try a 4-level Coif18 wavelet packet transform. Using the same quantizations as before—9 bpp for the trend and 6 bpp for the fluctuations—we obtain an estimated 0.49 bpp. That represents a 36% improvement over the 0.77 bpp estimated for the wavelet compression discussed in Section 2.9. In Table 4.3 we show a comparison of these two compressions of Fingerprint 1. Although the wavelet packet transform compression does not produce as small a relative 2-norm error as the wavelet transform compression, nevertheless, a value of 0.043 is still better than the 0.05 rule of thumb value for an acceptable approximation. In fact, the compressed version of Fingerprint 1 produced by the wavelet packet transform is virtually indistinguishable from the original (hence we do not include a figure of it).[2] Taking into account the other data from Table 4.3—the number of significant transform values and the number of bpps—it is clear that the wavelet packet compression of Fingerprint 1 is superior to the wavelet compression.

Table 4.3 Two compressions of Fingerprint 1

Transform	*Sig. values*	*Bpp*	*Rel. 2-norm error*
wavelet	33330	0.77	0.035
w. packet	20796	0.49	0.043

This last example gives us some further insight into the standard method adopted by the FBI for fingerprint compression, the WSQ method. The WSQ method achieves *at least* 15:1 compression, without noticeable loss of detail, on all fingerprint images. It achieves such a remarkable result by applying a hybrid wavelet transform compression that combines features of both wavelet and wavelet packet transforms. In this hybrid transform, not every subimage is subjected to a further 1-level wavelet transform, but a large percentage of subimages are further transformed. For an example illustrating why this might be advantageous, consider the two transforms of the *Barb* image in Figure 4.2. Notice that the vertical fluctuation $\mathbf{v}^1$ in the lower right quadrant of Figure 4.2(a) does not seem to be significantly compressed by applying another 1-level wavelet transform. Therefore, some advantage in compression might be obtained by *not* applying a 1-level wavelet transform to this subimage, while applying it to the other subimages. An exact description of the hybrid approach used by the WSQ method can be found in the articles listed in the Notes and references section for this chapter.

4.3 Continuous wavelet transforms

In this section and the next we shall describe the concept of a *continuous wavelet transform* (CWT), and how it can be approximated in a discrete

[2]You can, however, find the compressed image at the FAWAV website.

form using a computer. We begin our discussion by describing one type of CWT, known as the *Mexican hat* CWT, which has been used extensively in seismic analysis. In the next section we turn to a second type of CWT, the Gabor CWT, which has many applications to analyzing audio signals. Although we do not have space for a thorough treatment of CWTs, we can nevertheless introduce some of the essential ideas.

The notion of a CWT is founded upon many of the concepts that we introduced in our discussion of discrete wavelet analysis in Chapters 1 through 3, especially the ideas connected with discrete correlations and frequency analysis. A CWT provides a very redundant, but also very finely detailed, description of a signal in terms of both time and frequency. CWTs are particularly helpful in tackling problems involving signal identification and detection of hidden transients (hard to detect, short-lived elements of a signal).

To define a CWT we begin with a given function $\Psi(x)$, which is called the *analyzing wavelet.* For instance, if we define $\Psi(x)$ by

$$\Psi(x) = 2\pi w^{-1/2} \left[1 - 2\pi(x/w)^2 \right] e^{-\pi(x/w)^2}, \quad w = 1/16, \tag{4.7}$$

then this analyzing wavelet is called a *Mexican hat wavelet,* with *width parameter* $w = 1/16$. See Figure 4.3(a).

It is possible to choose other values for w besides $1/16$, but this one example should suffice. By graphing the Mexican hat wavelet using different values of w, it is easy to see why w is called a width parameter. The larger the value of w, the more the energy of $\Psi(x)$ is spread out over a larger region of the x-axis.

The Mexican hat wavelet is not the only kind of analyzing wavelet. In the next section, we shall consider the Gabor wavelet, which is very advantageous for analyzing recordings of speech or music. We begin in this section with the Mexican hat wavelet because it is somewhat easier to explain the concept of a CWT using this wavelet.

Given an analyzing wavelet $\Psi(x)$, then a CWT of a discrete signal $\mathbf{f}$ is defined by computing several correlations of this signal with discrete samplings of the functions $\Psi_s(x)$ defined by

$$\Psi_s(x) = \frac{1}{\sqrt{s}} \Psi\left(\frac{x}{s}\right), \quad s > 0. \tag{4.8}$$

The parameter s is called a *scale parameter.* If we sample each signal $\Psi_s(x)$ at discrete time values $t_1, t_2, \ldots, t_N$, where N is the length of $\mathbf{f}$, then we generate the discrete signals $\mathbf{g}_s$ defined by

$$\mathbf{g}_s = (\Psi_s(t_1), \Psi_s(t_2), \ldots, \Psi_s(t_N)).$$

A CWT of $\mathbf{f}$ then consists of a collection of discrete correlations $(\mathbf{f}: \mathbf{g}_s)$ over a finite collection of values of s. A common choice for these values is

$$s = 2^{-k/M}, \quad k = 0, 1, 2, \ldots, I \cdot M$$

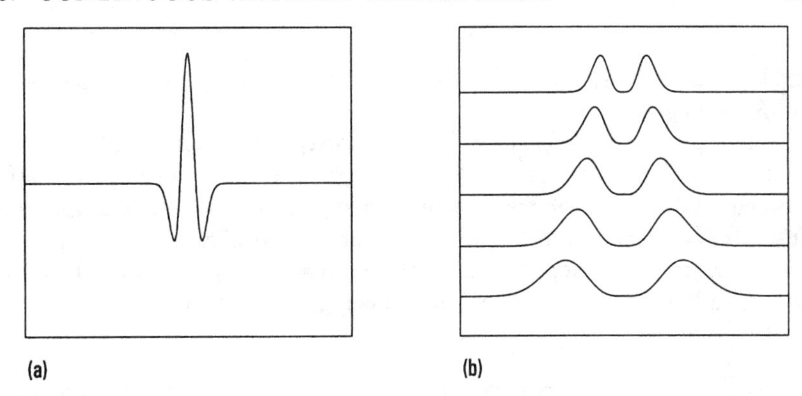

FIGURE 4.3
(a) The Mexican hat wavelet, $w = 1/16$. (b) DFTs of discrete samplings of this wavelet for scales $s = 2^{-k/6}$, from $k = 0$ at the top, then $k = 2$, then $k = 4$, then $k = 6$, down to $k = 8$ at the bottom.

where the positive integer I is called the number of *octaves* and the positive integer M is called the number of *voices* per octave. For example, 8 octaves and 16 voices per octave is the default choice in FAWAV. Another popular choice is 6 octaves and 12 voices per octave. This latter choice of scales corresponds—based on the relationship between scales and frequencies that we describe below—to the scale of notes on a piano (also known as the *well-tempered scale*).

At this point the reader may well be wondering what the point of all this is. One purpose of computing all these correlations that make up a CWT is that *a very finely detailed frequency analysis* of a signal can be carried out by making a judicious choice of the width parameter w and the number of octaves and voices. To see this, we observe that Formula (3.40) tells us that the DFTs of the correlations $(\mathbf{f}\!:\mathbf{g}_s)$ satisfy

$$(\mathbf{f}\!:\mathbf{g}_s) \overset{\mathcal{F}}{\longmapsto} \mathcal{F}\mathbf{f}\,\overline{\mathcal{F}\mathbf{g}_s}. \tag{4.9}$$

For a Mexican hat wavelet, $\mathcal{F}\mathbf{g}_s$ is real-valued; hence $\overline{\mathcal{F}\mathbf{g}_s} = \mathcal{F}\mathbf{g}_s$. Therefore Equation (4.9) becomes

$$(\mathbf{f}\!:\mathbf{g}_s) \overset{\mathcal{F}}{\longmapsto} \mathcal{F}\mathbf{f}\,\mathcal{F}\mathbf{g}_s. \tag{4.10}$$

Formula (4.10) is the basis for a very finely detailed frequency decomposition of a discrete signal $\mathbf{f}$. For example, in Figure 4.3(b) we show graphs of the DFTs $\mathcal{F}\mathbf{g}_s$ for the scale values $s = 2^{-k/6}$, with $k = 0$, 2, 4, 6, and 8. These graphs show that when these DFTs are multiplied with the DFT of $\mathbf{f}$, they provide a decomposition of $\mathcal{F}\mathbf{f}$ into a succession of finely resolved frequency bands. It should be noted that these successive bands overlap each other, and thus provide a very redundant decomposition of the DFT of $\mathbf{f}$. Notice also that the bands containing higher frequencies correspond to smaller scale values; *there is a reciprocal relationship between scale values and frequency values.*

A couple of examples should help to clarify these points. The first example we shall consider is a test case designed to illustrate the connection between a CWT and the frequencies in a signal. The second example is an illustration of how a CWT can be used for analyzing an ECG signal.

For our first example, we shall analyze a discrete signal $\mathbf{f}$, obtained from 2048 equally spaced samples of the following analog signal:

$$\begin{aligned} &\sin(40\pi x)e^{-100\pi(x-.2)^2} \\ &+ [\sin(40\pi x) + 2\cos(160\pi x)]\, e^{-50\pi(x-.5)^2} \\ &+ 2\sin(160\pi x)e^{-100\pi(x-.8)^2} \end{aligned} \tag{4.11}$$

over the interval $0 \leq x \leq 2$. See the top of Figure 4.4(a).

The signal in (4.11) consists of three terms. The first term contains a sine factor, $\sin(40\pi x)$, of frequency 20. Its other factor, $e^{-100\pi(x-.2)^2}$, serves as a damping factor which limits the energy of this term to a small interval centered on $x = 0.2$. This first term appears most prominently on the left-third of the graph at the top of Figure 4.4(a). Likewise, the third term contains a sine factor, $2\sin(160\pi x)$, of frequency 80, and this term appears most prominently on the right-third of the signal's graph. Notice that this frequency of 80 is four times as large as the first frequency of 20. Finally, the middle term

$$[\sin(40\pi x) + 2\cos(160\pi x)]\, e^{-50\pi(x-.5)^2}$$

has a factor containing both of these two frequencies, and can be observed most prominently within the middle of the signal's graph.

The CWT, also known as a *scalogram,* for this signal is shown at the bottom of Figure 4.4(a). The analyzing wavelet used to produce this CWT was a Mexican hat wavelet of width 1/16, with scales ranging over 8 octaves and 16 voices. The labels on the right side of the figure indicate *reciprocals* of the scales used. Because of the reciprocal relationship between scale and frequency noted above, this reciprocal-scale axis can also be viewed as a frequency axis. Notice that the four most prominent portions of this scalogram are aligned directly below the three most prominent parts of the signal. Of equal importance is the fact that these four portions of the scalogram are centered on two reciprocal-scales, $1/s \approx 2^{2.2}$ and $1/s \approx 2^{4.2}$. The second reciprocal scale is four times larger than the first reciprocal scale, just as the frequency 80 is four times larger than the frequency 20. Bearing this fact in mind, and recalling the alignment of the prominent regions of the scalogram with the three parts of the signal, we can see that the CWT provides us with a *time-frequency portrait* of the signal.

Although we have shown that it is possible to correctly interpret the meaning of this scalogram; nevertheless, we can produce a much simpler

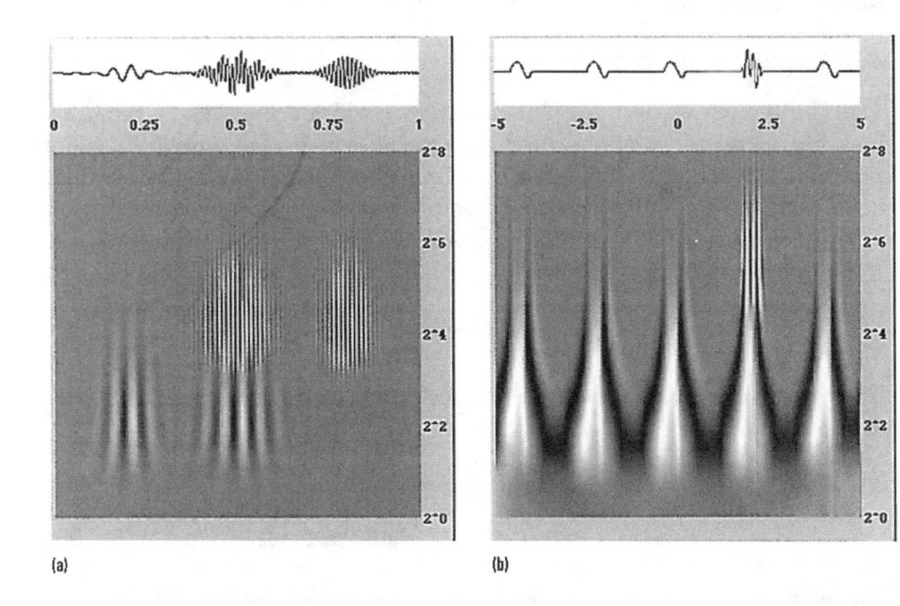

FIGURE 4.4
(a) Mexican hat CWT (scalogram) of a test signal with two main frequencies. (b) Mexican hat scalogram of simulated ECG signal. Whiter colors represent positive values, blacker values represent negative values, and the grey background represents zero values.

and more easily interpretable scalogram for this test signal using a Gabor analyzing wavelet. See Figure 4.5(a). We shall discuss this Gabor scalogram in the next section.

Our second example makes use of a Mexican hat CWT for analyzing a signal containing several transient bursts, a simulated ECG signal that we first considered in Section 3.4. See the top of Figure 4.4(b). The bottom of Figure 4.4(b) is a scalogram of this signal using a Mexican hat wavelet of width 2, over a range of 8 octaves and 16 voices. This scalogram shows how a Mexican hat wavelet can be used for detecting the onset and demise of each heartbeat. In particular, the aberrant, fourth heartbeat is singled out from the others by the longer vertical ridges extending upwards to the highest frequencies (at the eighth octave). Although this example is only a simulation, it does show the ease with which the Mexican hat CWT detects the presence of short-lived parts of a signal. Similar identifications of transient bursts are needed in seismology for the detection of earthquake tremors. Consequently, Mexican hat wavelets are widely used in seismology.

4.4 Gabor wavelets and speech analysis

In this section we describe Gabor wavelets, which are similar to the Mexican hat wavelets examined in the previous section, but provide a more powerful tool for analyzing speech and music. We shall first go over their definition, and then illustrate their use by examining a couple of examples.

A *Gabor wavelet,* with width parameter w and frequency parameter ν, is the following analyzing wavelet:

$$\Psi(x) = w^{-1/2} e^{-\pi(x/w)^2} e^{i2\pi\nu x/w}. \tag{4.12}$$

This wavelet is complex valued. Its real part $\Psi_{\mathrm{R}}(x)$ and imaginary part $\Psi_{\mathrm{I}}(x)$ are

$$\Psi_{\mathrm{R}}(x) = w^{-1/2} e^{-\pi(x/w)^2} \cos(2\pi\nu x/w), \tag{4.13a}$$

$$\Psi_{\mathrm{I}}(x) = w^{-1/2} e^{-\pi(x/w)^2} \sin(2\pi\nu x/w). \tag{4.13b}$$

The width parameter w plays the same role as for the Mexican hat wavelet; it controls the width of the region over which most of the energy of $\Psi(x)$ is concentrated. The frequency parameter ν provides the Gabor wavelet with an extra parameter for analysis.

One advantage that Gabor wavelets have when analyzing sound signals is that they contain factors of cosines and sines [see (4.13a) and (4.13b)]. These cosine and sine factors allow the Gabor wavelets to create easily interpretable scalograms of those signals which are combinations of cosines and sines—the most common instances of such signals are recorded music and speech. We shall see this in a moment, but first we need to say a little more about the CWT defined by a Gabor analyzing wavelet.

Because a Gabor wavelet is complex valued, it produces a complex-valued CWT. For many signals, it is often sufficient to just examine the magnitudes[3] of the Gabor CWT values. In particular, this is the case with the signals analyzed in the following two examples.

For our first example, we use a Gabor wavelet with width 1 and frequency 2 for analyzing the signal in (4.11). The graph of this signal is shown at the top of Figure 4.5(a). As we discussed in the previous section, this signal consists of three portions with associated frequencies of 20 and 80. The magnitudes for a Gabor scalogram of this signal, using 8 octaves and 16 voices, are graphed at the bottom of Figure 4.5(a). We see that this *magnitude-scalogram* consists of essentially just four prominent, and clearly

[3]Recall that a complex number z has a *magnitude* $|z|$ equal to its distance from the origin in the complex plane.

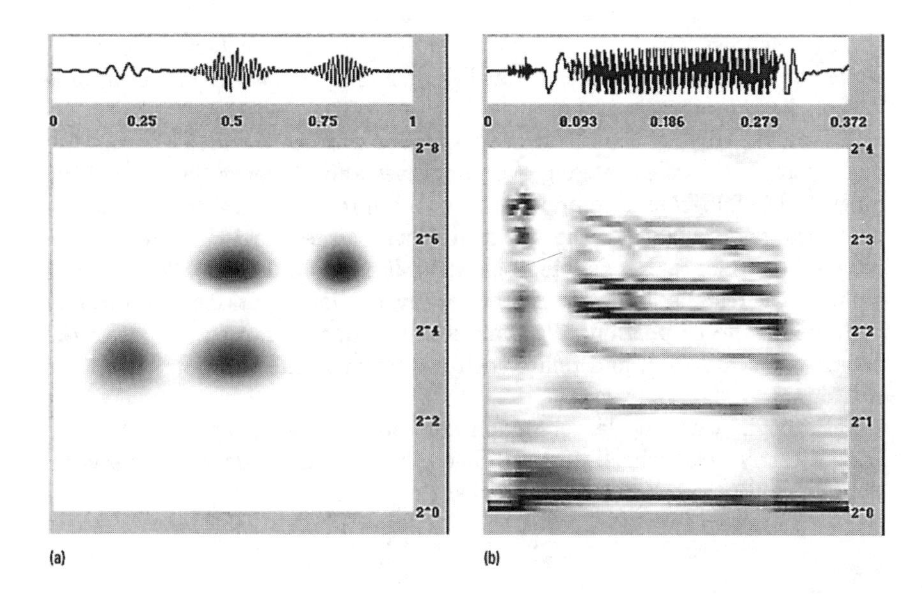

FIGURE 4.5
(a) Magnitudes of Gabor scalogram of test signal. (b) Magnitudes of Gabor scalogram of *call* sound. Darker regions denote larger magnitudes; lighter regions denote smaller magnitudes.

separated, spots aligned directly below the three most prominent portions of the signal. These four spots are centered on the two reciprocal-scale values of $2^{3.38}$ and $2^{5.38}$, which are in the same ratio as the two frequencies 20 and 80.

It is interesting to compare Figures 4.4(a) and 4.5(a). The simplicity of Figure 4.5(a) makes it much easier to interpret. The reason that the Gabor CWT is so clean and simple is because, for the proper choices of width w and frequency ν, the test signal in (4.11) consists of terms that are identical in form to one of the functions in (4.13a) or (4.13b). Therefore, when a scale value s produces a function $\Phi_{\rm R}(x/s)/\sqrt{s}$, or a function $\Phi_{\rm I}(x/s)/\sqrt{s}$, having a form similar to one of the terms in (4.11), then the correlation $(\mathbf{f} : \mathbf{g}_s)$ in the CWT will have some high-magnitude values.

This first example might appear to be rather limited in scope. After all, how many signals encountered in the real world are so nicely put together as this test signal? Our next example, however, shows that a Gabor CWT performs equally well in analyzing a real signal: a speech signal.

In Figure 4.5(b) we show a Gabor magnitude-scalogram of a recording of the author saying the word *call*. The recorded signal, which is shown at the top of the figure, consist of two main portions. These two portions correspond to the two sounds, *ca* and *ll*, that form the word *call*. The *ca* portion occupies a narrow area on the far left side of the *call* signal's graph, while the *ll* portion occupies a much larger area consisting of the middle half of the *call* signal's graph.

To analyze the *call* signal, we used a Gabor wavelet of width 1/8 and frequency 16, with scales ranging over 4 octaves and 16 voices. The resulting magnitude-scalogram is composed of two main regions lying directly underneath the two portions of the *call* signal. The largest region is a collection of several horizontal bands lying below the *ll* portion. The smaller region is a narrow, vertical segment consisting of several dark spots aligned directly underneath the *ca* portion. We shall now examine these two regions of the magnitude-scalogram, and relate their structure to the two portions of the *call* signal.

Let's begin with the larger region consisting of seven horizontal bands lying directly below the *ll* portion. These horizontal bands are centered on the following approximate reciprocal-scale values:

$$2^{0.17},\ 2^{1.17},\ 2^{1.7},\ 2^{2.17},\ 2^{2.5},\ 2^{2.97},\ 2^{3.17}. \tag{4.14}$$

If we divide each of these values by the smallest one, $2^{0.17}$, we get the following approximate ratios:

$$1,\ 2,\ 3,\ 4,\ 5,\ 7,\ 8. \tag{4.15}$$

Since reciprocal-scale values correspond to frequencies, we can see that these bands correspond to frequencies on a harmonic (musical) scale. In fact, in Figure 4.6(b) we show a graph of the spectrum[4] of a sound clip of the *ll* portion of the *call* signal. This spectrum shows that the frequencies of peak energy in the *ll* portion have the following approximate values:

$$140,\ 280,\ 420,\ 560,\ 700,\ 980,\ 1120. \tag{4.16}$$

Notice that these frequencies have the same ratios to the lowest frequency of 140 as the ratios in (4.15). There is even a missing frequency of $6 \times 140 = 840$, corresponding to a missing horizontal band in the magnitude-scalogram that appears to be centered along the reciprocal-scale $2^{2.7}$ (a small part of this missing band is visible below the right edge of the *ll* portion). In fact, the reciprocal-scale $2^{2.7}$ is about 6 times the lowest value of $2^{0.17}$.

This region illustrates an important property of many portions of speech signals, the property of *frequency banding*. These frequency bands are called

[4]The spectrum of a signal was discussed in Section 3.2.

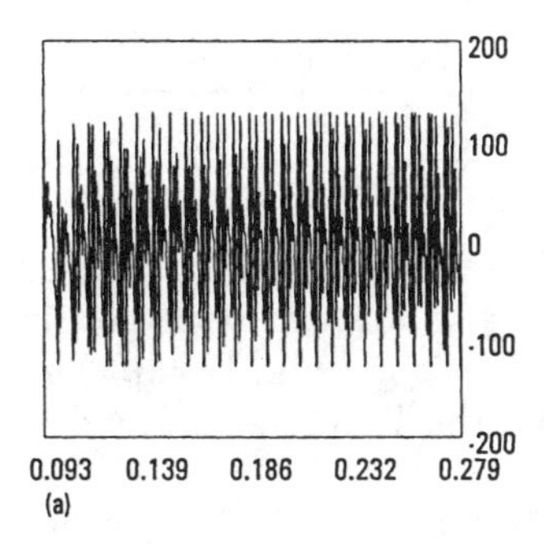

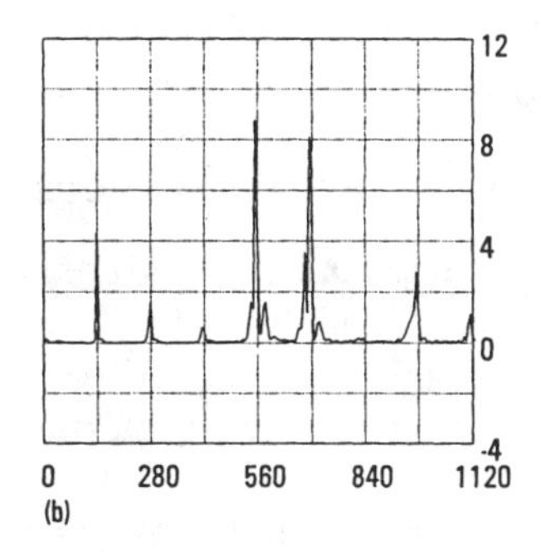

FIGURE 4.6
(a) A portion of the *ll* sound in the *call* signal; the horizontal axis is the time axis. (b) Spectrum of the signal in (a); the horizontal axis is the frequency axis.

formants in linguistics. All speakers, whether they are native speakers of English or not, produce a sequence of such frequency bands for the *ll* portion of *call*. For some speakers, the bands are horizontal, while for other speakers the bands are curved. The *ll* sound is a fundamental unit of English speech, called a *phoneme*.

The second region of the magnitude-scalogram lies below the *ca* portion. The *ca* sound is distinguished clearly from the *ll* portion by its lack of formants. From the magnitude-scalogram, we see that the *ca* portion is composed of a much more widely dispersed, almost continuous, range of frequencies without any significant banding.

This last example shows what a powerful tool the Gabor CWT provides for analyzing a speech signal. We were able to use it to clearly distinguish the two portions in the *call* sound, to understand the formant structure of the *ll* portion, and to determine that the *ca* portion lacks a formant structure.

Another application of these Gabor scalograms is that, when applied to recordings of different people saying *call*, they produce visibly different scalograms. These scalograms function as a kind of "fingerprint" for identifying different speakers. Furthermore, the ribboned structure of formants for the *ll* portion is displayed for all speakers, although they trace out different curves for different speakers. For the reader who wishes to verify these statements, we have included several recordings of different speakers saying the word *call* at the FAWAV website.

4.5 Notes and references

The best introductory material on wavelet packet transforms can be found in [WI1] and [CW1]. There is also a good discussion in [CW2]. A very thorough treatment of the subject is given in [WI2]. The relation between wavelet packet transforms and the WSQ method is described in [BBH].

Rigorous expositions of the complete theory of CWTs can be found in [DAU] and [LMR]. A more complete treatment of the discrete version described in this primer is given in [MAL]. For a discussion of the uses of the CWT for analysis of ECGs, see [STC]. Applying Gabor CWTs to the detection of engine malfunctions in Japanese automobiles is described in [KOB]. An interesting relationship between CWTs and human hearing, with applications to speech analysis, is described in [DAM]. Background on formants and phonemes in linguistics can be found in [ODA].

Appendix A

Software for wavelet analysis

> A computational study is unlikely to lead to real scientific progress unless the software environment is convenient enough to encourage one to vary parameters, modify the problem, play around.
>
> *Lloyd N. Trefethen*

> I don't think I could have done anything if I hadn't had a little computer and some graphics output.
>
> *Alex Grossmann*[1]

In order to apply wavelets to practical problems, a computer software environment is essential. The software FAWAV which produced the figures for this primer provides such an environment. FAWAV can be downloaded from the following website:

`http://www.crcpress.com/edp/download/`

free of charge. For simplicity, we shall refer to the web address above as the FAWAV website. When you visit the FAWAV website there will be instructions for how to proceed (see also Section A.2 below). It is planned that this site will be updated periodically with new versions of FAWAV, and with new data to supplement the ideas discussed in this primer. In this appendix, we begin by providing a brief overview of the essential features of FAWAV. We then provide some further details on installing FAWAV. At the end of the appendix, we list some of the other software that is available.

[1]Trefethen's quote is from [TRE]. Grossmann's quote is from [BUR].

A.1 Description of the book's software

FAWAV is designed to perform 1D and 2D wavelet analysis and Fourier analysis. It runs under WINDOWS™ 95, 98, and NT 4.0, providing a standard WINDOWS™ type of environment. In order to simplify the process of studying wavelet analysis, FAWAV is designed so that *it requires no programming on the user's part.* You simply graph functions—either by using formulas as in a graphing calculator, or by loading external audio files or image files—and then make menu selections and respond to dialog boxes in order to perform analysis on these signals. Besides the FAWAV program itself, the package that you initially download will include a User's Manual and extensive on-line help.

FAWAV can perform the following kinds of wavelet analysis:

- Wavelet transforms: Haar, Daub$J$, Coif$I$ transforms;
- Wavelet series: Haar, Daub$J$, and Coif$I$ series;
- Wavelet packet transforms: Walsh, Daub$J$, and Coif$I$ transforms;
- Wavelet packet series: Walsh, Daub$J$, and Coif$I$ series;
- Scalograms (1D only).

Except for scalograms, all of these procedures are capable of handling both 1D and 2D data.

Wavelet transforms, wavelet packet transforms

Wavelet transforms are easy to perform with FAWAV. After you have a signal displayed on screen, then you select *Transforms/Wavelet* from the FAWAV menu. A dialog box appears and you can select a type of wavelet to use—either Coif$I$, Daub$J$, or Haar—and specify the number of levels to use in the transform. Wavelet packet transforms are done in the same way. The only difference is that you begin by selecting *Wavelet packet* from the *Transform* menu.

Wavelet series, wavelet packet series

These procedures were not explicitly discussed in the primer, but were used for generating many of the examples. A *wavelet series* is just a convenient term for the following three step process:

1. Compute a transform, either wavelet or wavelet packet.

2. Modify the transform values from Step 1.

3. Compute the inverse transform of the modified values from Step 2.

There are a number of ways to carry out the modification of the wavelet transform values in Step 2. The most common method, which can be done in both 1D and 2D, is to threshold the values. When a threshold method is used, we say that the three step process produces a *thresholded series.* As with wavelet transforms, you produce wavelet series by some simple menu choices.

Scalograms

Scalograms are easy to compute with FAWAV. Once you have a signal displayed on screen, then you select *Analysis/Scalogram* from the menu. You then specify which analyzing wavelet you want to use, how many octaves and voices to use, etc. FAWAV will then plot the discrete scalogram. We used this procedure to plot the scalograms shown in Figures 4.4 and 4.5.

Fourier analysis

Besides wavelet analysis, FAWAV also provides several methods for performing Fourier analysis, some of which we made use of in Chapter 3. These methods are

- Fourier series;
- Fourier transforms (DFTs) and power spectra;
- Sine and cosine series and transforms (1D only);
- Convolutions, autocorrelations, and pair correlations;
- Spectrograms (1D only).

The methods that we used in the primer are DFTs, power spectra, and pair correlations. These procedures are all invoked with simple menu choices and dialog boxes.

Audio and image editors

FAWAV includes a simple audio editor, which can be used for loading and modifying `.wav` files. The main purpose of this editor is to allow you to clip portions out of sound files for analysis. This audio editor is only in a preliminary stage of development at this time. It is planned that future versions of FAWAV will contain a more robust editor with many more features.

Besides the audio editor, FAWAV includes procedures for image processing. At present, these procedures only handle special gray-scale test images of the kind described in the primer; but updates of FAWAV will extend these features in order to create a full-featured image processing environment within FAWAV.

A.2 Installing the book's software

To install FAWAV on your computer you must first download the compressed file `fawav.zip` from the FAWAV website's address given at the start of this Appendix. Once you have downloaded `fawav.zip`, you must then decompress it using PKUNZIP (other programs, such as WINZIP, may also be used for decompressing). After decompressing, you then run the program `setup.exe` and follow the instructions.

The setup program is a standard WINDOWS™ setup program, so we shall not describe it any further, except to say that as a safeguard it will not overwrite any existing files on your computer. Later, if you want to uninstall FAWAV, you can do so by selecting the *Add/Remove Programs* option from the WINDOWS™ *Control Panel*.

Once you finish installing FAWAV, you should look at the `read.me` file. It contains last-minute information about FAWAV. If you are interested in updates and/or future versions of FAWAV, you should periodically visit the FAWAV website.

In order to save room on your hard disk, this installation of FAWAV does *not* copy all of the data files—image files, audio files, figure files, etc.—from the FAWAV website to your hard disk. You should download these files separately as you need them.

A.3 Other software

FAWAV can only serve as an introductory tool for applications of wavelet analysis. It does not contain all of the huge number of various algorithms that have been created, nor does it include all of the manifold types of wavelets that exist. Fortunately, there are several excellent software packages available, many of them freely available for downloading from the

Internet. Here is a partial list:

- WAVELAB is a set of MATLAB™ routines developed by Donoho and his collaborators at Stanford. It can be downloaded from the following FTP site:

  `ftp://playfair.stanford.edu/pub/wavelab`

  or from the following website:

  `http://www.playfair.stanford.edu/~wavelab.`

- WAVELET TOOLBOX is a collection of MATLAB™ routines written by researchers at Rice University. It is available from the following website:

  `http://www-dsp.rice.edu/software/RWT.`

- LASTWAVE is a set of C language routines, including a command shell environment written by several French researchers. It can be obtained from the website:

  `http://www.cmap.polytechnique.fr/users/www.bacry.`

- AWA3 (Adapted Waveform Analysis Library, v. 3) is a collection of C language routines for wavelet analysis produced by *Fast Mathematical Algorithms and Hardware Corporation.* For further information, send an e-mail to the following address:

  `victor@math.wustl.edu.`

There are many other software packages available. The book [HUB] contains a listing of wavelet analysis software, as do [MAL] and [BGG]. A great deal of the latest information on wavelet analysis—including announcements of new software and the latest updates to existing software—can be obtained by signing on to the free electronic journal, *Wavelet Digest,* which can be done by accessing the following website:

`http://www.wavelet.org.`

References

[AKA] M. Akay. Diagnosis of Coronary Artery Disease Using Wavelet-Based Neural Networks. In [ALU], pages 513–526.

[ALU] A. Aldroubi, M. Unser, editors. *Wavelets in Medicine and Biology.* CRC Press, Boca Raton, FL, 1996.

[ASH] R.B. Ash. *Information Theory.* Dover, New York, NY, 1990.

[BAS] S.I. Baskakov. *Signals and Circuits.* Mir, Moscow, 1986.

[BBH] J.N. Bradley, C.M. Brislawn, T. Hopper. The FBI Wavelet/Scalar Quantization Standard for gray-scale fingerprint image compression. *SPIE,* Vol. 1961, *Visual Information Processing II (1993)*, pp. 293–304.

[BEF] J.J. Benedetto, M.W. Frazier, editors. *Wavelets. Mathematics and Applications.* CRC Press, Boca Raton, FL, 1994.

[BGG] C.S. Burrus, R.H. Gopinath, H. Guo. *Introduction to Wavelets and Wavelet Transforms, A Primer.* Prentice-Hall, Englewood Cliffs, NJ, 1998.

[BRH] W.L. Briggs, V.E. Henson. *The DFT. An Owner's Manual.* SIAM, Philadelphia, PA, 1995.

[BRI] E.O. Brigham. *The Fast Fourier Transform.* Prentice-Hall, Englewood Cliffs, NJ, 1978.

[BRS] C.M. Brislawn. Fingerprints Go Digital. *Notices of the Amer. Math. Soc.,* Nov. 1995.

[BUD] J.B. Buckheit, D.L. Donoho. WaveLab and reproducible research. In *Wavelets and Statistics,* pp. 53–81, Springer, Berlin, 1995. Edited by A. Antoniadis, G. Oppenheim.

[BUR] B. Burke. The Mathematical Microscope: waves, wavelets, and beyond. In *A Positron Named Priscilla, Scientific Discovery at the Frontier,* pages 196–235, National Academy Press, 1994. Edited by M. Bartusiak.

[CDL] A. Chambolle, R.A. DeVore, N-Y. Lee, B.J. Lucier. Nonlinear Wavelet Image Processing: Variational Problems, Compression, and Noise Removal through Wavelet Shrinkage. *IEEE Trans. on Image Proc.,* Vol. 7, No. 1, Jan. 1998.

[CHU] C.K. Chui. *Wavelets: A Mathematical Tool for Signal Analysis.* SIAM, Philadelphia, PA, 1997.

[COT] T.M. Cover, J.A. Thomas. *Elements of Information Theory.* Wiley, New York, NY, 1991.

[CW1] R.R. Coifman, M.V. Wickerhauser. Wavelets and Adapted Waveform Analysis. A Toolkit for Signal Processing and Numerical Analysis. In [DAE], pp. 119–154.

[CW2] R.R. Coifman, M.V. Wickerhauser. Wavelets and Adapted Waveform Analysis. In [BEF], pp. 399–424.

[DAE] I. Daubechies, editor. *Different Perspectives on Wavelets.* AMS, Providence, RI, 1993.

[DAM] I. Daubechies, S. Maes. A Nonlinear Squeezing of the Continuous Wavelet Transform Based on Auditory Nerve Models. In [ALU], pp. 527–546.

[DAU] I. Daubechies. *Ten Lectures on Wavelets.* SIAM, Philadelphia, PA, 1992.

[DJK] D. Donoho, I. Johnstone. Ideal spatial adaptation via wavelet shrinkage. *Biometrika,* Vol. 81, pp. 425–455, December 1994.

[DOJ] D. Donoho, I. Johnstone, G. Kerkyacharian, D. Picard. Wavelet shrinkage: asymptopia? *J. of Royal Stat. Soc. B.,* Vol. 57, No. 2, pp. 301–369, 1995.

[DON] D. Donoho. Nonlinear Wavelet Methods for Recovery of Signals, Densities, and Spectra from Indirect and Noisy Data. In [DAE], pp. 173–205.

[DVN] G.M. Davis, A. Nosratinia. Wavelet-based Image Coding: An Overview. *Applied and Computational Control, Signals and Circuits,* Vol. 1, No. 1, Spring 1998.

[DVS] G.M. Davis. A Wavelet-Based Analysis of Fractal Image Compression. *IEEE Trans. on Image Proc.,* Vol. 7, No. 2, Feb. 1998.

[FAN] R.L. Fante. *Signal Analysis and Estimation.* Wiley, New York, NY, 1988.

[FHV] M. Farge, J.C.R. Hunt, J.C. Vassilicos, editors. *Wavelets, Fractals and Fourier Transforms.* Clarendon Press, Oxford, 1993.

[FI1] D. Field. Scale Invariance and Self-Similar "Wavelet" Transforms: An Analysis of Natural Scenes and Mammalian Visual Systems. In [FHV], pp. 151–193.

[FI2] D. Field. What is the Goal of Sensory Coding? *Neural Computations,* Vol. 6, No. 4, 1994, pp. 559–601.

[HAM] R.W. Hamming. *Numerical Methods for Scientists and Engineers.* McGraw-Hill, New York, NY, 1962.

[HER] V.K. Heer, H-E Reinfelder. A comparison of reversible methods for data compression. In *Medical Imaging IV,* pages 354-365. Proceedings SPIE, No. 1233, 1990.

[HEW] E. Hernandez, G. Weiss. *A First Course on Wavelets.* CRC Press, Boca Raton, FL, 1996.

[HHJ] D. Hankerson, G.A. Harris, P.D. Johnson, Jr. *Introduction to Information Theory and Data Compression.* CRC Press, Boca Raton, FL, 1998.

[HUB] B. Burke Hubbard. *The World According to Wavelets, Second Edition.* AK Peters, Wellesley, MA, 1998.

[JAH] B. Jähne. *Digital Image Processing.* Springer, New York, NY, 1995.

[KOB] M. Kobayashi. Listening for Defects: Wavelet-Based Acoustical Signal Processing in Japan. *SIAM News,* Vol. 29, No. 2, March 1996.

[LMR] A.K. Louis, P. Maaß, A. Rieder. *Wavelets, Theory and Applications.* Wiley, New York, NY, 1997.

[MAL] S. Mallat. *A Wavelet Tour of Signal Processing.* Academic Press, New York, NY, 1998.

[ME1] Y. Meyer. *Wavelets and Operators.* Cambridge University Press, Cambridge, UK, 1992.

[ME2] Y. Meyer. *Wavelets. Algorithms and Applications.* SIAM, Philadelphia, PA, 1993.

[MLF] M. Malfait. Using Wavelets to Suppress Noise in Biomedical Images. In [ALU], pp. 191–208.

[ODA] W. O'Grady, M. Dobrovolsky, M. Arnoff. *Contemporary Linguistics, An Introduction.* St. Martins Press, New York, 1993.

[RAJ] M. Rabbini, P.W. Jones. *Digital Image Compression Techniques.* SPIE Press, Bellingham, WA, 1991.

[RAO] K.R. Rao. *Fast transforms: algorithms, analyses, applications.* Academic Press, New York, NY, 1982.

[REW] H.L. Resnikoff, R.O. Wells, Jr. *Wavelet Analysis. The Scalable Structure of Information.* Springer, New York, NY, 1998.

[RUS] J.C. Russ. *The Image Processing Handbook.* CRC Press, Boca Raton, FL, 1995.

[SAP] A. Said, W.A. Pearlman. A new, fast, efficient image codec based on set partitioning in hierarchical trees. *IEEE Trans. on Circuits and Systems for Video Tech.,* Vol. 6, No. 3, pp. 243–250, June 1996.

[SHA] J.M. Shapiro. Embedded image coding using zerotrees of wavelet coefficients. *IEEE Trans. on Signal Proc.,* Vol. 41, No. 12, pp. 3445–3462, December 1993.

[STC] L. Senhadji, L. Thoraval, G. Carrault. Continuous Wavelet Transform: ECG Recognition Based on Phase and Modulus Representations and Hidden Markov Models. In [ALU], pp. 439–464.

[STN] G. Strang, T. Nguyen. *Wavelets and Filter Banks.* Wellesley-Cambridge Press, Boston, 1996.

[STR] R.S. Strichartz. How to Make Wavelets. *The American Math. Monthly,* Vol. 100, No. 6, June–July, 1993.

[SW1] W. Sweldens. The Lifting Scheme: A Construction of Second Generation Wavelets. Technical Report TR-1995-6, Math. Dept. Univ. of South Carolina, May 1995.

[SW2] W. Sweldens. The lifting scheme: a custom-design construction of biorthogonal wavelets. *Applied and Computational Harmonic Analysis,* Vol. 3, No. 2, pp. 186–200, 1996.

[TRE] L.N. Trefethen. Maxims About Numerical Mathematics, Computers, Science, and Life. *SIAM News,* Vol. 31, No. 1, 1998.

[VEK] M. Vetterli, J. Kovačević. *Wavelets and Subband Coding.* Prentice-Hall, Englewood Cliffs, NJ, 1995.

[WA1] J.S. Walker. *Fourier Analysis.* Oxford, New York, NY, 1988.

[WA2] J.S. Walker. *Fast Fourier Transforms, Second Edition.* CRC Press, Boca Raton, FL, 1996.

[WA3] J.S. Walker. Fourier Analysis and Wavelet Analysis. *Notices of the Amer. Math. Soc.,* Vol. 44, No. 6, June–July, 1997, pp. 658–670.

[WAN] B.A. Wandell. *Foundations of Vision.* Sinauer Associates, Sunderland, MA, 1995.

[WAT] A.B. Watson. Efficiency of a model human image code. *J. Optical Soc. Am.,* Vol. 4, No. 12, December 1987, pp. 2401–2417.

[WI1] M.V. Wickerhauser. Best-adapted Wavelet Packet Bases. In [DAE], pp. 155–172.

[WI2] M.V. Wickerhauser. *Adapted Wavelet Analysis from Theory to Software.* A.K. Peters, Wellesley, MA, 1994.

[XWH] Y. Xu, J.B. Weaver, D.M. Healy, J. Lu. Wavelet transform domain filters: a spatially selective noise filtration technique. *IEEE Trans. Image Processing,* Vol. 3, No. 6, 1994, pp. 747–758.

Index